Executive Safety and International Terrorism

A Guide for Travellers

Executive Safety and International Terrorism

A Guide for Travellers

Anthony J. Scotti

PRENTICE-HALL, INC., Englewood Cliffs, New Jersey 07632

Editorial/production supervision and
interior design: LISA SCHULZ
Cover design: LUNDGREN GRAPHICS, LTD.
Manufacturing buyer: CAROL BYSTROM

The publisher offers discounts on this book when ordered
in bulk quantities. For more information, write:

Special Sales/College Marketing
Prentice-Hall, Inc.
College Technical and Reference Division
Englewood Cliffs, NJ 07632

Printed in the United States of America

10 9 8 7 6 5 4 3 2 1

ISBN 0-13-294380-8 025

Prentice-Hall International (UK) Limited, *London*
Prentice-Hall of Australia Pty. Limited, *Sydney*
Prentice-Hall Canada Inc., *Toronto*
Prentice-Hall Hispanoamericana, S.A., *Mexico*
Prentice-Hall of India Private Limited, *New Delhi*
Prentice-Hall of Japan, Inc., *Tokyo*
Prentice-Hall of Southeast Asia Pte. Ltd., *Singapore*
Editora Prentice-Hall do Brasil, Ltda., *Rio de Janeiro*

Contents

Foreword

Recent world events have compelled us to confront a new brand of fear. This fear is strong enough to empty the seats aboard transatlantic flights and to transform overseas vacations into cross-country family outings. How much do we really know and understand about terrorism? We're all familiar enough with the *aftermath* of terrorist incidents: for a decade television networks have been delivering into our living rooms images of American hostages and rubble-filled Middle Eastern cities. Most importantly, what steps can we take to *avoid* becoming the victim of a terrorist attack? How much do we know about the *goals* and *methods* of terrorists? Are their selections arbitrary or planned?

Executive Safety and International Terrorism is the first book written for all of us—**executives, diplomats, vacationers, and other international travellers.** As Anthony Scotti points out, everyone is vulnerable to the increasing threat of world terrorism. Airline tickets and boarding passes do not identify us as "executive," "vacationer," or "CEO." Once aboard an aircraft, terrorists cannot distinguish tourists from executive vice-presidents—forcing us all to exercise extreme caution.

In light of this, Anthony Scotti goes to great lengths to address the needs and concerns of all international voyagers. *The result is a book filled with precautions and safety tips that are as relevant to the tourist as they are to the executive vice-president.* In addressing the needs of these non-executive groups, Scotti never loses sight of his main objective—that of educating **all travellers** on how to protect themselves from the threat of terrorism.

While screaming headlines and zealous news reporters have heightened our awareness of world terrorism, this attention has done little to provide useful insights into the motives underlying most terrorist movements. Given the amount of media attention and coverage terrorism receives, it is appalling just how little we comprehend the true nature of terrorism. Anthony Scotti is convinced that much of our fear stems from our lack of understanding. One of his goals in writing this book is to increase our knowledge, and thereby reduce our fear.

An important lesson presented by Scotti is that terrorists are not arbitrary or random; their operations are planned and directed, and are often as precise as a military drill. Once we understand this, we can protect ourselves against the threat of terrorism—not with guns and weapons, but with common sense and measured vigilance.

Unlike other books on the subject, *Executive Safety and International Terrorism* is not a political treatise that has emerged from the proceedings of an international symposium. This book does not contain lengthy tirades expounding upon the political implications of terrorism or discourses weighing the corroding effects of terrorism on East-West relations. What you *will* find are useful insights and suggestions which can help diminish the likelihood that either you or someone close to you will be involved in a terrorist incident.

Jeffrey A. Krames
Publisher

Preface

We are living in the age of terrorism and it shows no signs of going away. International terrorism is a growth industry, one that's shifting its sights towards business.

The book is about protecting you, your family, and the company you own or work for against the ravages of terrorism. It's for those who have gained prominence in their field of business. It's for managers of multinationals who are living in or about to move to a country with a terrorism problem. It's for anyone who feels they might become a terrorist target.

The basic problems with terrorism are that no one seems to understand it or know how to effectively guard against it. This book explains the effects of terrorism on the day-to-day operations of business overseas, along with what you can do to make yourself a less tempting target for terrorism.

This book shows how terrorists affect the bottom line. While fighting terrorism has become a major cost factor in doing business abroad, terrorism costs businesses more than money; it takes its toll on the mental and physical well-being of executives and their families. With this book, readers will learn how to protect themselves while at home, at work, and in transit. In short, this book will explain how to transform yourself into an unattractive target for terrorism, with a focus on avoiding kidnappings. To do this, we'll look at kidnapping through the terrorists' eyes, learning how terrorists select their targets. Why is one person chosen as a kidnap victim over another? Once you understand why terrorists strike where they do, you'll use this information to develop your own personal security program.

Terrorism can and will strike anywhere, at any time. This book covers terrorism's most common battle grounds, teaching you how to develop a specific defense for each major trouble area. Hand-in-hand with anti-terrorism defense goes the potential use of force to protect you and your family. The use of lethal and non-lethal force is a controversial issue, one that is covered extensively in this book. Should you decide to use force, we'll examine the options available to you, along with the pro's and con's of each option.

You'll soon realize that when living in a terrorist environment, you should be ready for the worst possible situation: being kidnapped. We'll show you how to set up an emergency kidnap plan for yourself and your family. If you are kidnapped, you'll be put through a dangerous and devastating ordeal. The book covers what to do at the moment of capture, the most dangerous and deadly part of the crime of kidnapping. Your blunders at the wrong moments could provoke a violent reaction by terrorists that could easily result in serious injury—or worse.

Once kidnapped, you'll need to know how to survive captivity. This book discusses what to do while being held captive, right up to the point of what to do if you are about to be rescued. Although being rescued sounds like a happy ending to the torment of captivity, it can actually be far more dangerous than the kidnapping itself.

The problems of skyjacking receive thorough coverage, along with what to do and how to act while being held hostage aboard an aircraft.

In summary, the book shows you why, as an overseas executive, you have become a target for terrorism in recent years, and explains how to detect terrorists planning a move against you. You'll learn how to make yourself less of a target, and if you do get kidnapped, what to do to enhance your chance of survival while in captivity. You'll learn how to survive in today's dangerous world.

Anthony J. Scotti

Acknowledgments

I would like to acknowledge the people that made this book possible.

Joyce Huber, who got the book off the ground, and with her pleasant persistence, pushed the book along.

Jeff Krames, for feeling that I could write the book.

Cathy McMahon and Laurie Nazzaro for spending weeks in front of a word processor, typing, organizing, and praying over the book.

My wife Judy, and daughter Toni-Ann, for having patience while I wrote the book, and for all those people that I have encountered all over the world that have indirectly contributed to this book. May it make all of us just a little bit safer.

Executive Safety and International Terrorism

A Guide for Travellers

1

The Effects of Terrorism on Business

INTRODUCTION TO TERRORISM

Terrorism is the universally recognized crime, a crime that at a moment's notice can reach into every facet of our lives, and can, at times, dominate our existence. Not long ago terrorism was an issue that did not concern the people of the United States, its businesses, or its businessmen. It was something that happened to other people in far-away places.

In the past, the first thought that came to mind when someone mentioned terrorism was that of the ongoing Arab-Israeli conflict. Today, the situation is entirely different. Terrorism is no longer locked away in some obscure corner of the globe. From the boardrooms of multinational corporations to the farmlands of Middle America, Americans are witnesses to and subjects of the violence and mayhem of terrorism. Nearly every day the media brings us pictures of American travelers held hostage at gunpoint, unfortunate citizens whose only mistake was to be in the wrong place at the wrong time. The news tells us about businessmen kidnapped and held captive for no reason other than they are American businessmen. On any given night the six o'clock news brings pictures of buildings in rubble into our living rooms, and the heart-breaking sight of flag-draped coffins being carried from an airplane; often the results of a lone madman driving a truckload of bombs, all in the name of terrorism.

With just one incident, terrorists can control worldwide media. Through modern communications, a hijacking or hostage situation can be presented to viewers around the world as it is happening. A small band of people can easily seize the attention of every American, exploiting the opportunity to detail their grievances. Television enables hundreds of millions of people to watch terrorism's mayhem and destruction. In part, it is this ability to gain the world's attention that makes terrorism work. Most terrorist attacks are designed as media events. The bombings, hijackings, and kidnappings are intended to get the attention of both the electronic media and international press. Terrorists want to be on the six o'clock news. Terrorism is theater, and as Americans, we are becoming the actors. Terrorism is widespread and indiscriminate. Terrorists attack anything and anybody, hijacking airliners, trains, even ocean liners. They kidnap priests and nuns. They blow up nightclubs, department stores, churches, and computers. Nothing is spared.

Eventually terrorism strikes close to home. Someone from your home town, someone who went to your school or works for your company can be the victim of terrorism. For an example of how terrorism reaches the heartland of the U.S., we only have to look at the town of Wausau, Wisconsin, population 32,000. This small Middle American town lost two of its native sons to terrorist acts at opposite ends of the world in a space of six months.

First, Wausau native Charles Hegna was murdered in an Iranian airport following the hijacking of a Kuwaiti airliner. Six months later another native of Wausau, Marine Sgt. Patrick Kwiatkowski, was gunned down in El Salvador when terrorists sprayed an outdoor cafe with gunfire. In the space of just a few months, terrorists had twice brought grief, anger, and frustration to a small American town.

Despite an increased awareness of terrorism the average American does not understand it. There are so many unanswered questions. Americans ask "Why is so much hate and anger directed towards us?" "What did we do to these people that makes them hate us?" "Why do they select Americans as targets for their hate and anger?" Terrorism's complexity makes these all hard questions to answer, partly because we generally see only a small portion and particular type of terrorism, the type directed against large groups of people, such as the passengers of a jetliner. Such major terrorist incidents make headlines and attract our attention. In between these terrorist "spectaculars," we tend to ignore the routine background of bombings, extortions, and kidnappings of individuals. However, it is precisely these type of actions that should alert us to the fact that terrorism is waging a continuous war against the Western nations. We must understand that we are in a continuous war against terrorism, and that there are day to day movements in this war that could, in the long run, prove to be devastating to the economic foundations of the U.S.A.

Part of this movement is an organized effort against a particular target: the businessman. Today, more than ever, businessmen are the terrorists' preferred target. People in business everywhere ask the same question: Why are terrorists selecting them as targets? In the business community, where people seek and need hard facts to make decisions, terrorism is an enigma.

Before the events of the past ten years, terrorism wasn't something that much concerned the business community. Terrorist attacks against businessmen were considered rare, never gaining much attention in or out of the business community. But terror attacks against business are not new. Businesspeople throughout the world have been targets of terrorism since the early 60's. It has been during the last decade that terrorism has become a major problem for corporations both large and small in the U.S. and overseas. As governments spend more money attempting to defend against terrorism, and strive to develop a hard line against the terrorist, terrorism has shifted its tactics towards attacking business and businesspeople. A study conducted by Risk International indicates that in the first half of 1985, 35% of all terrorist attacks were directed against business.

TERRORISM DEFINED

If terrorism is not hard to define, terrorism's targets and applications are. When terrorist tactics are applied to business, they nearly defy explanation. Business shouldn't be embarrassed by its inability to understand or explain terrorism. The U.N. has never gotten beyond theoretical explanations of terrorism, and more than likely never will. Whether an act of violence is terrorism, general crime, sabotage, or part of a war of independence depends on the social perspective of the individual. In point of fact, very few people are willing to call themselves terrorists. Terrorism has no precise definition. But a common general definition is that of an act of violence or the threat of violence coupled with an intention to create fear.

The individual who commits the terrorist act is often easier to define. Most terrorists take credit for what they have done. In many cases they will contact a news service or prominent journalist to explain their cause and why they committed an act of terrorism. Criminals do not usually put an ad in the paper taking credit for their acts.

The need for proponents of a particular cause to draw the world's attention to injustice, either real or imagined, is the major reason for terrorism. Many feel that terrorism is caused by misery, oppression, hunger, or other situations that otherwise powerless peoples often find themselves in.

But this definition does not make sense when terrorism becomes widespread in countries like Great Britain, Spain, Germany, and Italy. In these nations, terrorism often has a two-pronged purpose: combining political protest with a way for terrorists to raise funds through either kidnapping or outright theft. In the western democracies, terrorism's role as a way for the disenfranchised to gain a forum for their cause takes on a secondary role, superseded by the terrorists' needs to gut the soft underbelly of capitalism in order to get the funds they need. The intensity of violence employed by terrorists in these countries is no less than that of third world terrorism.

If we are to swallow the popular definition of terrorism, we must assume that terrorists are only the poor, the hungry, or the oppressed. However, history indicates this is not an accurate picture. Moreover, it makes no sense. The best way of eliminating misery is to supply the populace with a place to work, and a way of earning a decent living. When a business enters the community or, in the case of multi-nationals, enters a foreign country it is doing just that; supplying people with a place to work, and a chance to raise their standard of living. When a terrorist group bombs the workplace, or kidnaps a manager, it is doing nothing to solve the people's problems. Terrorists will invariably employ some sort of rhetoric that justifies their actions in the name of the people, when in fact terrorists attack the very system that supplies the people with work. When a terrorist organization concentrates on attacking American businesses and businessmen, it is trying to force American business interests out of their country, causing jobs to be lost and services for their people to be diminished.

This simply makes no sense. More often than not, when we take a closer look at terrorist acts, they all seem senseless and irrational. We should not be fooled! Terrorism is not senseless, and the terrorists are not mindless. As we will learn later, they have goals and objectives.

KNOWING THE ENEMY

Most businessmen have a distorted view of who the terrorists are. More often than not, terrorists are not the poor, or the oppressed. Many terrorists are middle or upper class idealists, who for some reason or other feel they can change the world for the better by getting rid of the capitalists. These terrorists usually come from good families, with parents that are doctors, lawyers, engineers, or some other kind of professionals.

Terrorists are usually well educated. The world's most notorious terrorist is a man who calls himself "Carlos the Jackal," taking his name from the title character in Frederick Forsythe's novel, *The Day of the Jackal.* Carlos's real name is Illyich Raminez Sanches. Born a Venezuelan, he comes from a very comfortable background. His father is a millionaire, and his mother is a socialite divorced from her husband, living in London. Educated in Paris and Moscow, he is neither poor, hungry, nor oppressed.

A composite portrait of a terrorist would reveal a person with an average age of 22 to 24 years, well-educated, from an upper middle-class family. The type of education this person receives is what often creates the problem. Like the Jackal, the educational backgrounds of most terrorists are Marxist. This leads them to consider American firms as the center of all they believe to be wrong with society. Throughout their education they are trained to think of the corporation and its executives as a single entity. Since capitalism causes all the problems in the world, the people operating capitalistic businesses are therefore responsible for the problems. That is one kind of terrorist. There is another.

They are the religious fanatics. Basically assassins, their methods date back to the days of the Crusades. They will give up their lives to kill the enemy, for they are assured they will meet their God if they die killing his enemies. Impossible to negotiate with because they often have no demands, all these fanatics want to do is kill. They don't care who, where, or when.

The political terrorist is not impossibly difficult to defend against. The religious terrorist is nearly impossible to defend against.

Businessmen should not be misled by what they may read or hear about terrorism in the media. There are times when well meaning representatives try to offer a "justified explanation" of terrorist actions. For example, in a speech presented at the U.N., Yasir Arafat was quick to point out that in 1776 George Washington would have been a terrorist in the eyes of the British. Arafat was insinuating that if a group is fighting for freedom from an oppressive government, they are freedom fighters, not terrorists. As the saying goes, "One man's terrorist is another man's freedom fighter."

This is patently absurd. To mention the word "freedom" in the same breath as the word "terrorist" is obscene. It is relatively easy to distinguish between freedom fighters and terrorists. Freedom fighters don't kill innocent women and children; freedom fighters don't assassinate innocent businessmen; freedom fighters don't hijack airliners and hold innocent women and men hostage while they perform for the media. Unfortunately for all of us, the word "terrorist" has come to mean something less than "criminal." All terrorists are criminals, but all criminals are not terrorists.

As far as business is concerned, a terrorist is a person who commits an act of violence, be it sabotage, murder, or kidnapping, against the corporation's facilities or personnel. Although terrorists will attack buildings and other inanimate objects, their acts of violence will more often take the form of kidnapping or extortion against the individual. By extorting corporations through kidnapping or threats, terrorists are trying to force them to finance terrorist activities, intimidate employees, or to exploit the companies' status to gain publicity for their cause. Left unchecked, terrorism forces corporations to change the way they do business in many parts of the world. The changes ultimately, and adversely, affect the corporations' bottom lines.

Not all terrorists are looking for money. As mentioned before, terrorism is theater, a way to gain attention and public sympathy for causes. Over the years, this terrorist method of going to the public has proven successful. These methods are deplorable but unfortunately effective. Terrorism simply works.

As an executive of an American company doing business overseas you will have to protect and prevent against the violence of terrorism. To do this you will have to understand unconventional warfare, and learn about terrorist weapons and tactics. First, learn to identify your enemy. Who are the terrorists? Where do they operate? What do they want? Once more, terrorists can be separated into different categories.

DIFFERENT TYPES OF TERRORISTS

Political terrorists: A term generally applied to those individuals whose motivation, at least publicly, is avowed to be ideological in nature. The methods used to achieve their ends are usually ruthless, and the victims of their violence need have no connection with the target. If you are not with them, you are against them, and therefore the enemy. Part of their philosophy seems to be that the more heinous the act, the more effective it is in achieving the goal. This group will machinegun passengers waiting to get on an airliner, or blow up airplanes with innocent people on board. This group seeks to establish what they feel to be their legitimate right to lead the existing government. You do not have to be a direct enemy to be a target. If you fly on the enemy's airline, work in the enemy's country, do business with them, then you are the enemy too. This type of terrorist is deadly, but inconsistent. Many times they will not interfere with certain businesses, depending on those businesses' areas of operation—perhaps realizing that if they do eventually take over the government that expelling the companies that bring their natural resources to the world marketplace would be counterproductive in the long run. This philosophy holds true in some parts of the world but not others. In many areas of the Middle East the mere accident of being an American makes you a target. However, not being in the Middle East does not exempt anyone from being caught up in indiscriminate and random violence.

The religious fanatic is an offshoot of the political terrorist. They have few goals other than to inflict as much damage and pain as possible. Their targets are selected to cause the most damage. They are extremely dangerous to everyone, not just businesspeople. Sometimes their tactics are so deranged they become dangerous to their own people. Keep in mind that the group that recently opened fire on passengers waiting to get on an El Al Airline in Rome had the goal of killing as many children as possible.

Criminal terrorists or bandits: Are defined as those who perpetrate the same act as terrorists but whose motivations are solely monetary. In certain areas of the world, specifically South and Central America, it is becoming difficult to tell the difference between the political terrorists and the financially motivated criminals. It is sometimes easier for corporations to meet the demands of criminals than to satisfy the political purposes behind terrorists' demands. In Italy common criminals have emulated the terrorist tactics with great success. Engaging in illegal activities for profit, employing terror whenever necessary; this group is motivated simply by money. They may camouflage their goals with political rhetoric but what they really want is loot. Despite a lack of political motivation, this group is no less dangerous to the businessman. Businesses and businessmen are the ones that have the money and over the years have shown a willingness to pay terrorists what they want. In Italy, most of the kidnappings have nothing to do with terrorism. In that country, criminals are using terrorist tactics as a moneymaking venture.

Insane terrorists: Individuals who are usually set into motion by extensive publicity for someone else's terrorist act, a desire to get that kind of notoriety themselves, or by a grudge against their targets. Due in part to the publicity afforded the exploits of both terrorists and criminals, deranged terrorists represent a continuing threat. With notable exceptions, they rarely combine the careful planning and execution found in other groups, engaging instead in acts of terrorism as the result of serious personal, phychological disorders. Because of this underlying irrationality, their actions may be even more dangerous because they are so very unpredictable. In the business community, they are often disgruntled employees who consider the executive as an extension of the company that caused their problems. The answer? Get even with the company! They will then vent their hostilities on the executives. It is easy to confuse a person who commits an act of random violence or someone who is insane with a band of politically-committed terrorists who will give up their lives for a cause. Although the act may appear to be that of an insane man, it may be that of a political terrorist.

Average citizens turned terrorists: This group rejects peaceful methods of settling differences, instead using violence to express their rage and dramatize their demands. Sometimes these groups preach peace but do it violently. A good example of these groups are some anti-nuclear, anti-abortion, and animal rights groups. Anti-abortion groups have blown up clinics, and threatened the life of doctors. Animal rights advocates have broken into medical labs, often destroying experiments and freeing dangerous test animals. Members of some anti-nuclear groups have broken into airbases and other military facilities in order to disrupt activities and destroy military equipment.

It is that any given terrorist group is a melange of people from all of the groups above. Experience has shown that a terrorist group will use criminals to meet certain goals and criminals will use a terrorist group to meet certain financial goals.

TERRORIST AREAS OF OPERATION

Political terrorist groups can be broken down into the geographical areas in which they operate.

Transnational groups: These groups basically act on their own and although they may not be controlled by one government, they receive financial assistance and arms from one or more governments. This is also called "State Supported Terrorism." Regimes such as Libya, the Soviet Union, North Korea, Vietnam, and Syria will supply and train such groups. Although all these countries deny the fact that terrorist training camps exist within their borders, there is information to prove them wrong. While transnational groups operate in more than one area, their allegiance is to "the cause" and to their group. They are highly motivated and well-trained in modern terrorist tactics. Their intent is to produce havoc, cause destruction, and spread fear. They travel from country to country with the help of their sponsoring nation, or nations. This type of group will walk into an airport and open fire on passengers who are waiting to get on an airline. The group that recently fired upon passengers in Italy waiting to get on an El Al airliner were not from Italy. They traveled to Italy to commit their act of terrorism.

International groups: International groups are similar to transnational groups in that they operate without regard to national boundaries; however, international groups are controlled or directed by specific sovereign governments. They may be used as surrogate armies by governments wanting supremacy without legal accountability. The Sandinistas in Nicaragua were and are supported by the Soviet Union and Cuba. There are groups in Africa and Central America supported by other governments. The Cubans are often active in many countries supplying training and equipment to terrorists.

National groups: National or domestic groups are autonomous and operate strictly within one country; more often than not, the country's standing government is not under the control or direction of the group. In Peru a group called the Shining Path operates within that country trying to overthrow the government. Copycat phenomenon is also encountered in Peru. Criminal groups are imitating terrorists; kidnapping a business and holding it for ransom.

TERRORIST GOALS

The goals of a terrorist or terrorist group may be either immediate or long range.

Immediate Goals

- *Obtain worldwide, national, or local recognition for "the cause."* Since the American media is the world's most powerful, kidnapping Americans and holding them hostage is equivalent to getting millions of dollars of free publicity. By now we've all seen how easy it is for terrorists to hold the United States helpless by taking hostages.
- *Force governments to overreact and create repression that would lead to public dissension.* Many governments respond inappropriately to terrorism. Governments can create curfews, impose work restrictions, or enforce frequently Draconian measures on corporations that make it difficult to conduct business as usual.
- *Expose a government's inability to protect its citizens.* If it is difficult or dangerous to get to work, and while at work the employees are worried about what is happening to their homes, the combined stresses make for low productivity, while eroding public confidence in both business and the government.
- *Obtain money or equipment.* There have been many American companies in all parts of the world which have contributed money to terrorist groups by paying ransom to retrieve their executives. When a terrorist group kidnaps an executive and the company pays a ransom to have him released, it is supplying terrorists with the ready cash needed to complete their objectives.
- *Discourage impending foreign investments or foreign government assistance programs.* Not too many companies are willing to invest in a country that has a terrorist problem.

CORPORATE LEGAL RESPONSIBILITY

Corporations agree they have a moral and legal responsibility to do all they can to protect facilities and personnel against acts of terrorism. Unfortunately, corporations find it easier to protect facilities than personnel. Property is easier to protect because the best equipment available uses state of the art technology. People are a lot easier to damage than facilities, however, and protecting a vulnerable individual requires more than equipment. Moreover, personnel are much more lucrative targets than buildings and equipment. For one thing, you can't easily hold a building hostage, so when terrorists want money to support their cause they kidnap corporate personnel.

Kidnapping is an old and simple method of extracting money from those who have it. Kidnapping brings misery to both the victim's family and the company. To all the hardships kidnapping causes to corporations we can now add a new one. Recent history indicates that when an employee of an American corporation is kidnapped abroad, it can present a bewildering series of legal problems to the corporation; not to mention some fairly basic and pretty important personal safety concerns for the person kidnapped.

Until recently, very few kidnappings involving personnel from U.S. corporations wound up in litigation. All that has changed dramatically over the last few years. The chances of a lawsuit being presented against the corporation by either kidnap victims or their families is now very high. Whether the kidnap victim is rescued and returned unharmed has no bearing on whether there will be a lawsuit.

There have been recent cases where kidnapped employees of multinational companies and their families have sued the company for not supplying proper protection against terrorism. A classic case of an employee who felt that he was not adequately protected is that of Gustavo Curtis.

Curtis was kidnapped while working for Beatrice Food in Bogota, Colombia. He was held for eight months while the company negotiated his release. The ransom payment was about $500,000. After his release, Curtis informed his employer that he was not willing to go back to work in Bogota, Colombia. His superiors informed him they expected him to return to Bogota. When he refused, they terminated his employment. On March 3, 1978, Curtis filed suit against Beatrice Food. In the suit, he claimed that Beatrice Food had not carried out its duty to protect him, and had neglected its duty to rescue him once he was kidnapped. The final outcome of the case was that Beatrice Food was found innocent of any wrong doing. The Gustavo Curtis case was dismissed because the judge felt that Curtis failed to assure his own personal safety.

In fact, Beatrice Food had done a good job of supplying Curtis with the necessary tools to develop an excellent personal security program. The company supplied him with seminars on how to reduce the risk of kidnappings. He had a security consultant at his disposal. The court found that he took no steps to implement security plans. Curtis had the authority to do whatever he felt necessary to protect himself, yet did nothing.

It is very important to note that Curtis was aware that he was a potential target. He had received warnings from the U.S. Embassy. He made the same mistake many other executives make: he did not take the problem of terrorism seriously.

In a terrorist environment you must make use of all the resources available to protect yourself. Most corporations are more than willing to supply you with everything needed to protect yourself. It is to their advantage to do so. From the corporation's viewpoint, it knows that it must supply you with all that is needed to protect yourself or face the possibility of a lawsuit if you are kidnapped. What's more, lawsuits are not just limited to victims of kidnapping.

On February 27th, 1976, William Niehous, a vice-president for Owens-Illinois in Caracas, Venezuela, was kidnapped from his home and held captive for an amazing 40 months. Quite by accident, he was rescued by the police, but before the rescue his wife filed a $4 million dollar lawsuit against Owens-Illinois for mishandling the negotiations. The suit was dropped after the rescue. The Niehous case showed that a spouse can lose confidence in the company's ability to handle the negotiations of a kidnapping while talks are still in progress, and then sue the company for the way it is handling the problem.

The Neihous case shows the need for corporations to develop a crisis management team for these sort of problems. It is imperative that companies working in a terrorist environment develop such teams. If a company employee is kidnapped, the company must do everything possible to get that employee back. It is important that companies organize their recovery efforts before the fact. If the company waits until the incident presents itself, the shock can create confusion along with unnecessary and unsafe delays, resulting in an executive that is severely injured, or worse, while in captivity.

Working for the Goodyear company in December of 1980, Clifford Bevens was kidnapped in Guatemala. Tragically, Bevens was shot and killed during a rescue attempt in August of 1981. Mrs. Bevens filed a wrongful death suit against Goodyear and received a $1.25 million dollar settlement from Goodyear. What was interesting about the case is that both Mrs. Bevens and Goodyear asked the court to seal the record. *The Wall Street Journal* quoted Goodyear's position. "It was, and remains, Goodyear's opinion that making the settlement public would serve to encourage terrorists and that refusal to meet their demands would only result in demands in U.S. courts by survivors. Further publicizing a settlement, whether large or small, would have established a price on our employees' heads."

The litigation problem does not end here. Once a company has paid ransom to get back a kidnapped executive, it can be sued for paying it. In 1981 Exxon Corp. was sued by one of its stockholders for paying $14.2

million in ransom for the release of Victor Samuelson, general manager of their refinery in Argentina. The court turned down the lawsuit.

Basically, the courts are saying that as long as the company gives an individual control of their security, and advises them of the potential danger of the assignment, it has fulfilled its obligation.

The company is not just obligated to warn its employees. Courts are also ruling that companies need to supply employees with the proper training and security equipment. To meet this need, the company should conduct security seminars for employees in countries that have terrorist problems, and provide managers with constant updates on the terrorist situation in these countries.

Don't be worried about alarming employees. It is far better to warn too much than not warn at all. An issue in the Curtis case was that Beatrice Food had done a good job of both warning Curtis and supplied him with the necessary equipment and training to protect himself. Beatrice Food supplied Curtis with lectures on how to avoid terrorist kidnappings by experts in the field. Therefore, the three issues a company has to think about before sending employees off to a terrorist environment are information, training, and equipment. Staying informed on terrorist situations throughout the world can be expensive and time-consuming. To keep abreast of this problem, many multinational companies have security personnel each responsible for a particular part of the world. Services by security professionals constantly evaluating the terrorist problems in a particular part of the world and developing prevention programs that supply employers with updated information are available. Good security personnel also offer training on how to avoid terrorist kidnappings.

The next part of the security equation is the proper protection equipment. This equipment must be supplied to you not only at the workplace, but also in the home. Even though the work day is completed, companies are responsible for the overseas protection of their employees. However, you must make a mental note not to rely solely on equipment to protect you. In order for personal protection to work, you must have everything going for you. Above all you have to develop the proper attitude towards personal security.

THE COST OF TERRORISM

A quarterly report produced by Risk International, a leading company in the field of risk analysis, indicates that there were 859 major terrorist incidents reported during April-June 1985 versus 676 incidents in the preceding quarter.

The loss to business worldwide from bombings alone was $88 million in 1981 and another $100 million in 1982. Moreover, these numbers are not as accurate as they could be. Many losses are not reported, so the amounts lost to bombings are probably much higher. Losses due to bombings and other terrorist attacks-coupled with the cost of preventing terrorism-affect the corporation's profit and loss statement. While important, it's not just a matter of dollars lost. The many ways terrorism affects employees cannot be accounted for in a profit or loss statement. Damage to the human factor can be devastating, and not just to the direct victims of terrorist action. A business must look at the effects a bombing, or kidnapping has on the morale of everyone working for the company. A few random acts of terrorism can cause a company to come to a standstill.

When terrorists decide to attack facilities, they often do not merely bomb buildings but can sometimes be adept at destroying the precise portion of the facility that will hurt the company most. For example, in recognition of the importance of computers to a company, terrorists often attack corporate computer facilities. Several companies doing business in foreign countries have reported this type of attack. Imagine the costs involved and the chaos created if a major corporation lost the use of its computer for just a few days.

Another hidden cost of terrorism is the effect that it has on employees' personal lives and psychological well-being. If a company is constantly harrassed by terrorists, workers lose confidence in the company's ability to protect them. This takes a toll on productivity. If you or your employees are under the constant threat of kidnapping you have to change the way you work. To protect yourself from kidnapping, you will have to incorporate changes in your work habits that may not be in the best interests of the company or you. Changing your way of living to compensate for terrorism can have a tremendously negative impact on you and your family. Terrorist activity can grow to the point that it is unsafe to go to any public gathering. This can wipe out your family's social life. Most executives in your position are tough-minded people who will do whatever it takes to get the job done. But if that means your family must live like hermits or soldiers, it may be better to send the family home. This is often the safer thing to do but it can carry a terrible social and personal price for both you and your family in terms of sheer separation, loneliness, and alienation.

The obvious conclusion is that guarding against terrorism is costly, both in terms of money and personal loss. Accordingly, the average American corporation is concerned about security and crime and the burden they bring to their personnel. Responsible corporations are doing something about it. In a study called *The Figgie Report* executives from Fortune 1000 companies were asked about the types of security their companies used:

88% have building security checks
66% have burglar alarms
48% have closed circuit T.V.

38% have electronic card identification systems
24% have armed guards

According to the American Management Association, crime against business is well over the $40 billion mark. By 1990, U.S. business will be spending more than $21 billion on private security.

Today, only a small part of the security budget is spent on protecting the executive. Executive protection is relatively new, and has created a new growth industry. Its basic objective is to stop the most devastating and expensive crime a company can experience, executive kidnapping. For the terrorist, kidnapping is good business. Throughout the years terrorists have realized that kidnapping executives is profitable and, for the most part, not a risky venture. Depending on terrorists'. goals, they will select one of two groups of people as targets for kidnapping: business executives or diplomats. Kidnapping diplomats gains publicity and has proven to be an effective means of releasing terrorists' comrades from prisons, but over the last few years many countries have adopted a policy of non-negotiation with terrorists. Not many governments are willing to spend money to get their diplomats back from the hands of terrorists, not because of the cost but because they feel that giving in to terrorist demands leads to more kidnappings. This "no negotiating policy" has made businessmen very lucrative targets. Unlike the diplomatic community, multinational companies are more willing to and capable of paying ransoms. Corporations in the past have paid ransoms bordering on the incredible. The largest ransom ever paid for the release of businessmen was for the Born brothers, owners of Argentina's largest private firm. The company paid a staggering *$60 million* for their return. The brothers were kidnapped by a terrorist group called "The Monteneros." The Monteneros were so bold they actually held a press conference a few days before releasing the brothers to explain how they pulled off the kidnapping and the terms of the ransom. The press conference looked like any big company's annual meeting, at which the president explains the details of a major acquisition to the stockholders. The Montenoros were so good at what they did that at one time they had a bank roll of $240 million, ready cash to support their activities. Although the Born brothers were the subject of the highest ransom paid there have been others that were nearly as spectacular.

Firestone paid $3 million in Argentina.
Exxon paid $14.2 million in Argentina.
Goodyear paid $10 million in Guatemela.
Amoco paid $3 million in Argentina.

Sometimes money is not the only issue. American firms have been used as pawns to get political exposure by the terrorist.

Beckman Instruments had a manager kidnapped in El Salvador. Their "payment" was to publish ads in major newspapers throughout the world calling for the overthrow of the El Salvadoran regime. Paying for ads that criticize the government in power does not do good things for your relationship with the host government.

Any company interested in profit has to protect its assets. Of all the assets that need protecting, there are none more important than the human asset. As an executive, you are one of your company's greatest assets, but in a terrorist environment you can become your company's greatest liability. Terrorists know the company will be willing and able to pay ransom to get you back. Kidnapping you is profitable, and the likelihood of your kidnapper being caught slim. Statisics indicate that terror against business is crime that really does pay, and is much safer than most other forms of terror. According to a study by the Conference Board, a business research organization, more than half of the 567 Americans kidnapped overseas between 1970 and 1978 were businessmen. Terror against American business is simply cost effective. Another study on the effectiveness of terrorist kidnapping conducted by the Rand Corp. think tank indicated a virtual 100% probability of gaining publicity through kidnapping if publicity was an original goal. Kidnapping an American businessman gives the terrorist access not only to the corporate bank account but to America's powerful media as well.

Startling statistics from the Rand Corp.'s study showed that a terrorist has a:

87% probability of success

83% probability of safe passage or exit for the terrorists or others if that was a sole demand

79% probability that all members of the terrorist team would escape punishment or death whether they seized hostages or not

69% probability that all or part of the demands beyond exit of safe passage would be met

67% probability that if no concessions were obtained through the action, virtually all the terrorists would escape, either by going underground, accepting safe passage, or by surrendering to a sympathetic government.

The other immeasurable cost of doing business in a terrorist environment is the loss of productivity due to low morale. It is hard for anyone to concentrate on business when they are in fear of their lives, and the lives of their families. Another element of the bottom line companies must face in dealing with terrorism is the simple fact that companies have a legal and moral responsibility to protect employees from violence.

HOW TERRORISTS SEE YOU

The person best equipped to protect you against terrorism is you. Surprised? Don't be. The personal battle against terrorism is not as difficult as you may imagine. The first step is learning how to assess the threat that terrorism poses to you. Just how vulnerable are you? Is there an immediate threat to you or your company?

You must be concerned about your own personal vulnerability. By far the biggest threat against you is kidnapping. In executive kidnapping the important issue is who the terrorists select as the target for a kidnapping and why. Terrorists have a wide number of options available to them. They can select people from the military, they can select diplomatic personnel, they can select businessmen or any of a myriad of influential and vulnerable decision makers.

As mentioned before, terrorists kidnap businessmen for money and/or to gain publicity for their cause. The question you must answer is simple: of all the target businessmen available, why select you?

You will attract terrorists' attention if you fit into what is called the **terrorist profile**. Understanding this profile is critical. You may be selected as a target if terrorists feel you fit the following criteria:

APPARENT WEALTH: Most terrorist groups need money. If you appear wealthy you will become a target. This wealth does not have to be your wealth. If you work for a big American company with "deep pockets," the apparent ability to pay a ransom, you have a good chance of being selected as a potential target.

REPRESENT SOMETHING IMPORTANT: Terrorists need publicity. Terrorists need to advertise just like businesses. It is vital for them to keep their cause in the eye of the public. Kidnapping someone unimportant does not get them the publicity they need. As an American manager working in a foreign land you represent something important. Your kidnapping will be newsworthy in itself. If you represent a prominent American firm with a worldwide image, such as a major multinational, your vulnerability is increased.

PARTICULARLY VALUABLE TO SOMEONE: You may feel that you are personally not worth the effort of a kidnapping. But you do have a value to the company you work for. The more valuable you are to your company the more likely you will be selected as a target.

ACCESSIBILITY: Terrorists are like everyone else. Terrorists seek the path of the least resistance. The easier you are to kidnap and the more accessible you are the more likely you will become a target.

When assessing your vulnerability to terrorism take a close look at yourself.*

1. Do you appear wealthy? Like beauty, wealth is in the eye of the beholder. Therefore, appearances count. Look at yourself through your kidnapper's eyes, colored by the social mores and economic standards of the country you are in. Do you live in a large residence? Do you have more than one house? Do you employ servants? Do you seem to be enjoying an exciting and expensive lifestyle? This appearance of wealth is very important. Realistically, any American enjoying what would be considered a modest standard of living in the Third World is living in comparative luxury, measured against local standards.
2. Are you of value to somebody who can and will do a great deal to secure your release? You may not be rich yourself, but you may be worth a great deal to someone else. As a manager of an American firm, you have intrinsic value. History has proven that American corporations will pay large ransoms to get you back.
3. If you are kidnapped will it provide publicity favorable to the kidnapper's cause? As an American working overseas your kidnapping will be news.

Again, the important issue is how terrorists see you. Step outside of the role of an American manager and look at yourself objectively. The average

*"Kidnapping: How To Avoid It, How To Survive It" *Clandestine Tactics and Technology: Tactics And Countermeasures,* (Gaithersburg, MD: The International Association of Chiefs of Police, Inc., 1979) pp. 7

manager working overseas probably enjoys a lifestyle that could not be reproduced in the U.S., at least not nearly as cheaply. In many South American and Middle Eastern countries it is not uncommon to find corporate managers living in expensive homes located in the most affluent sections of the city, catered to by servants, and chauffeured to and from work. These trappings would be a sign of wealth in any country. It's certainly not the lifestyle of the average Third World citizen. By local standards, you are certain to be considered wealthy.

Will your kidnapping get media attention? There is no question that your kidnapping would be a media event in the country where you were kidnapped. You are an influential person. The people who work for your company, along with the people that depend on your products or services, look at you as an influential person. As far as the workers are concerned, you make decisions that can change their lives and in some cases change the economy of the country. The American manager of a multinational company doing business in a foreign land is inherently important, influential, and newsworthy.

The bottom line is simple. It makes no difference whether a businessman really fits into one of the above categories. All that counts is that criminals or terrorists think he does!

Businesses are vulnerable because of their basic structure. All businesses are interested in profitable growth. Any security measure a multinational takes must be measured against profit and loss. The effectiveness of a personal security program is hard to measure in terms of dollars and cents. If a company spends $1 million on security and nothing gets damaged or stolen, and no one is kidnapped, some short-sighted executives will consider the money wasted. If an anti-kidnapping program is working, management may get the impression that because no one has been kidnapped the program is not needed.

This is ridiculous. Examine the alternatives. It's a little late to decide you need a personal security program after you have been kidnapped. The problem with security is that if it's working, it looks as though it's not needed.

There's no question there are situations where the cost of security can outweigh the loss. But that generally is true only if the loss is measured in dollars. There are costs associated with terrorism and violence that are difficult or impossible to measure if the only criteria is profit and loss. William Niehous, who was kidnapped in Venezuela, spent 40 months in the hands of the terrorists. What price would you put on 40 months in captivity? What's the dollar value in terms of your own personal trauma and the anguish your family would endure?

ASSESSING THE THREAT

For many companies the issues of terrorism and security are unpleasant topics they'd rather not discuss. Talking about terrorism is like talking about death or taxes, not a very pleasant subject. The worst thing you can do is refuse to recognize the danger. Adopt this attitude, and you increase the possibility of being selected as a target. There is no need for any American working overseas to exaggerate the danger, but a realistic risk appraisal can give you peace of mind while allowing you time to put the proper defense mechanisms into place. There is no need to become excessively paranoid, but in a potentially dangerous situation a little paranoia can keep you alive. As the old saying goes, "Just because you're paranoid doesn't mean they're not really after you."

You must evaluate the problem. Make a realistic risk assessment of your situation. *Don't sweep the issue under the carpet.* Executives sometimes justify this course of action on the grounds that "it won't happen to me, or if they are after me there's nothing that will stop them, or I cannot operate efficiently if my hands are tied with security precautions."

Security practitioners around the world agree that the greatest single factor in reducing your risk to terrorism is for you to take active security precautions. To have terrorists see you taking active, effective precautions is probably the most important single thing you can do. Neither political terrorists nor criminals can afford a fiasco. They generally strike when the situation offers guaranteed success, so they always seek the easiest target available. Businessmen who don't take precautions are soft, easy targets. The more visible your personal security precautions, the more likely the terrorists are to abandon the attack. From the terrorist point of view: "there's always another manager." Terrorists are probably not interested in you personally. They are far more interested in what you represent, and in most foreign countries there are many easy targets available. The more daunting you appear, the less likely you are to become a target.

There is documented proof for this type of terrorist deterrence. Two brothers named Hosein were once convicted of a major kidnapping in the United Kingdom. Ironically, the brothers didn't kidnap the person they really wanted. Security made them settle for second-best. For their original target they had selected the wife of the wealthy and famous publisher Rupert Murdoch. Because of Mrs. Murdoch's squad of personal security guards and their high state of readiness, the brothers concluded she would be too hard to kidnap. They moved on to another easier but less wealthy target. They were willing to accept less money for more chances of success. Since they were eventually convicted, it would seem they got neither. However, one point is clear: Mrs. Murdoch was not touched.

The same was true of Aldo Moro, a former president of Italy. He was not the Red Brigade's original target. The Brigade's original intention was the kidnapping of two other men. But the Red Brigade decided that both men were too heavily guarded so they went after Moro, the easier target. Terrorists take risks but like any good soldier, they take calculated risks. You must calculate your own risks in dealing with them. Accept the fact that you have a problem and confront it as you would any other. Start out by knowing who you're dealing with.

WHO ARE THE TERRORISTS?

One of the biggest errors a company can make is to underestimate terrorist capabilities. Terrorists are not a disorganized rabble. They are organized to a degree that would be admired by most of the companies and managers they plan to attack. They must be well organized. Terrorist cells are usually a small group of people who realize that their opponents, either multinationals or governments, are bigger, more powerful, and have far greater resources than they do. They must beat that company by being better prepared.

Likewise, a company organized against terrorism is a formidable deterrent for the terrorist. Some executives might scoff at this, but a company that uses its resources intelligently and efficiently to combat terrorism can make life very difficult for the terrorists, forcing them underground to seek another target. The information-gathering abilities of large corporations is a powerful weapon in the fight against terrorism. A forward-looking company can organize an intelligence-gathering network against terrorism in much the same way as it would organize against or analyze any other challenge it faces.

INFORMATION-GATHERING TECHNIQUES

Knowledge is truly power. In order to gather as much information as possible about the threat facing you and your company, ask your managers for their assessment of the local terrorist situation. Rely especially on your "old hands," experienced personnel with a long history "in country." The information they give can help in determining who the terrorists are, the overall level of local terrorist activity, and what other companies are doing to handle the problem. You will probably discover that the person with the best idea of what is going on within the country is your security manager. Most likely, he has your answers. If your company has no security manager, get advice from other American companies in the area that do have an on-site security specialist.

UNDERSTANDING THE ENEMY

A fundamental part of any threat analysis program is a grassroots understanding of the enemy. It takes a terrorist to get a terrorist. Start thinking like a terrorist. One of the best ways to start is by reading *The Mini-Manual of the Urban Guerrilla.* Used by terrorists worldwide as the bible for their battles, the *Mini-Manual* is the accepted textbook for urban guerrilla warfare, and can be found in terrorist training camps and safe houses the world over. One of the more frightening features of this manual is that it clearly targets American industry as an enemy, together with the people who represent that industry.

Carlos Marighella, author of *The Mini-Manual of the Urban Guerrilla* is a Brazilian terrorist who was killed in the late 1960s. While the name is hardly a household word, Marighella is the father of modern terrorism. If he were alive today, Marighella would be very proud of the work his students have been carrying on around the world. The fact that terror attacks at opposite ends of the world are often virtually identical in terms of weapons and tactics is a tribute to Marighella's work and the level of acceptance his work has received. Imitation, after all, is the sincerest form of flattery.

In *The Mini-Manual of the Urban Guerrilla,* Marighella describes step by step how to organize a campaign of terror against North American business executives and the North American business community. To quote from his manual:

> "Terrorism is an action, usually involving the placement of a bomb or fire explosion of great destructive power, which is capable of effecting irreparable loss against the enemy. . . . Although terrorism generally involves an explosion, there are cases in which it may also be carried out by the execution and systematic burning of installations and properties of North American companies."

Marighella describes kidnappings as:

> "The capturing and holding in a secret spot a police agent, a North American Spy, a political personality, or notorious enemies of the revolution. Kidnapping is used to exchange or liberate revolutionary comrades, or to force suspension of torture in the jail cells of the military dictatorship."
>
> "The kidnapping of North American residents or visitors constitutes a form of protection against the penetration of United States imperialism. Execution is the killing of a North American spy, or an agent of the dictatorship of a police torture."

Clearly, Marighella's book spotlights terrorists' targets, zeroing in on the North American businessman as the enemy.

USING THE MEDIA FOR INTELLIGENCE GATHERING

One of the most easily accessible sources of information is the local media. Local newspapers, magazines, radio, and TV are good, timely sources of information on current security conditions in the country. The attitude of the press towards terrorism is a good indication of the severity of the problem you will have to face. If the press seems to favor what the terrorists are doing, does not condemn the terrorist, and/or constantly portrays the American business community as the villain, then you can be assured the problem is severe. The type of newspapers you read is also important. Communist newspapers are something most of us are not used to in America, although they do exist in the U.S. Naturally, most of these papers consider everything American corporations do as wrong. In many countries both English and local-language papers can be found. You should read both. Reading as many newspapers as you can will give you a valuable perspective on the situation in the country, and is an invaluable tool in learning the local language, something you must also do.

If you have a problem understanding the language of the country, and cannot read the local newspapers, find the local English language publications. Allow them to tide you over while you learn the local language. Read, listen, and watch for all the local information you can get. If you're busy, assign someone from the office staff you can trust to be your media/intelligence person. This person can summarize important events and give you a daily report. Most people in your position are very busy, so having someone do the media gathering for you is a real time-saver as well as probably the only way you will ever get it done. However, don't leave all the information gathering to someone else. If you do, you will lose your feel for the country and the sixth sense you are trying to develop that will serve you as an alarm to danger.

One reliable method of intelligence gathering doesn't take much time and is actually pleasant work. You should have a shortwave radio receiver capable of picking up both Voice of America and BBC broadcasts. Not only is the programming aired by these organizations timely, accurate, and valuable, but these two radio networks offer interesting information on both the world situation and the ways it will influence your own small corner of the globe.

Both BBC and VOA broadcast in English and local languages. VOA provides a type of broadcast in what it calls "Special English" that can be a trifle tedious to listen to, designed as it is for those just learning English. The announcers speak slowly and use very simple words. Broadcasts in the local language can prove valuable in learning the local tongue, just as reading local newspapers are.

A dependable shortwave radio can be extremely valuable in the event of serious problems, such as a government takeover or coup détat of the government. If this happens, your only reliable source of information may be from outside the country.

Along with the ability to listen to the outside world, you should be able to contact people outside the country by means other than the phone system.

Telex is fast and reaches just about anywhere. Unfortunately, in the event of a national emergency, it is usually one of the first communications links to be severed. Communications can be made much easier if the country allows amateur (ham) radio operators. Ham radio operators can easily make contact with people thousands of miles away. Make the effort to find a ham operator, perhaps someone within your own company. If you don't have an amateur radio operator in the company, seek some (preferably Americans), and make arrangements to communicate with them in the event of an emergency. Most ham operators in the U.S. can connect their radios to the American telephone system, so a local ham operator could make it possible for you to phone home when no other system works.

While it's important to read the local newspapers, which allows you to know what's going on around you in your country, you should also have the capability of knowing what's going on outside the country and have the ability to contact people on the outside. Shortwave and ham radios have proven their value in many overseas crisis situations.

While information received from the local media can be a good source of information, keep in mind that you should judge local media by the conditions under which it operates. Raised in the U.S., most of us feel that all we read, hear, or see in the media is the truth. In many countries there is partial and, in many cases, complete control of the media. You must make a candid assessment of the level of "freedom of the press" in the country you're working in. If you have a hard time "reading between the lines," confidentially consult your local American consul. If there is censorship, try to judge how much it distorts the accuracy of the information and just how much and what type of information is being censored. This won't be easy. If there are terrorist incidents going on all around you and they are not reported in the media that is an indication that the country does not want to admit they have a terrorist problem. A country that does not want to admit a high level of terrorist activity has a big problem. If the media is under government control, you must assume that much of what you hear and read is inaccurate, that you are hearing only what the government wants you to hear. Unlike the U.S., where we can trust the media, for the most part, the overseas press can sometimes be guilty of overstatements, understatements, and other distortions, including outright lies, depending on their feelings towards the U.S. Be realistic about what you read, see, and hear. Before making a commitment based on information received from the media, the accuracy of the information should be cross-checked by other sources.

USING PEOPLE FOR INTELLIGENCE GATHERING

Friends and associates can be good sources of information. Unfortunately they can also be sources of inaccurate information. When it comes to rumors, living overseas is no different than living in the U.S. Rumors can take an insignificant incident and change it into the beginning of World War Three. Treat the information you receive as you would any other information—its value depends on its source. The quality of the information received is only as good as the individual providing it.

There are no hard and fast rules about judging the quality of information received. If information is judged by the source it comes from, then the information should come from an unimpeachable source, someone you respect, someone who is honest, but above all someone who can and will say "I don't know" when they really don't know. Look for someone with a great deal of experience in the country. Look for someone logical, rational, and realistic. In your search you will encounter people who are either paranoid or overconfident. These two extremes will destroy any security plan, and should be avoided. Someone who sees a terrorist behind every tree is of no use to you. Likewise, a person who believes that everything is coming up roses when that is obviously not the case is also someone to avoid. While it's not a bad idea to have open discussions with the people you select on terrorist problems, security matters that pertain to you and your family must be kept confidential. Never discuss the details of the company's personal security plan with anyone not directly connected with the plan, no matter how much their opinion is valued. That knowledge must remain with you, your family, and only the smallest possible number of company security personnel.

There are two basic types of information you will receive from your friends and associates. The first will be direct answers to the questions you are asking. Once you have established your information source as someone you respect, a person who meets all the criteria mentioned previously, ask questions that relate to important issues such as safe areas to live, persons or areas to avoid, reliable and discreet suppliers of security equipment, the reliability of local police, specific problems regarding local crime, safe schools for your children, individuals that can and cannot be trusted, suspected terrorists, suspected terrorist plans and areas of operation, and so forth. These are important questions that must be immediately answered upon arrival at a foreign destination, and these questions deserve the most accurate answers you can find. Much of the worry can be taken out of this process if your company has a security department to take care of it for you. The only thing you then have to do is listen and take their recommendations. The security department is there to help you make these decisions. Unlike other departments within the company, the security department should and will get involved with your personal life, helping you locate a secure home, a safe school for your kids, arrange for personal security as needed, etc. When entering a new country, a well-run, on-the-ball corporate security department can remove a tremendous burden from your shoulders.

The second source of information is the type that you don't ask for, usually provided courtesy of the local "rumor mill." Most people are acquainted with rumor mills as they exist in every neighborhood, club, and business in the United States. To paraphrase Churchill, never have so few contributed to misinforming so many, so often, so completely, and for so long. The reasons for these rumor mills can be attributed to the American imagination and our national ability to take elements of dull truth and transform them into breathtaking fiction.

Despite their innocent nature, rumor mills can play a large role in undermining your security. As the famous World War Two poster warned, "Loose lips sink ships." Rumor mills go into high gear when the local security situation begins to deteriorate or at least looks like it's deteriorating. The result is that as tensions increase, so does an abundance of erroneous information that's there for anyone who will listen. This information will almost always present a distorted picture of the situation: a picture either vastly exaggerated or completely glossed over. Use the rumor mill for what it is, a source of a vast amount of highly questionable information, but information that can tell you much about how people are thinking, if not what's actually happening. The mill is a random, but occasionally valuable source.

While rumors make for good gossip on the cocktail circuit, they have no place whatsoever in basic security planning. If what is heard is interesting, then verify or negate the story by checking with reliable sources. Here is another situation where you should take full advantage of your security department. Tell them what you heard and let them look for the grain of truth in it.

THREAT ASSESSMENT

Once you have amassed all this information, your next task is analysis. Again, if you have a security department, it can take a tremendous burden off your shoulders. But not everyone traveling abroad or living in a foreign land has the luxury of a security department to take care of their security needs. If you have to develop your own threat assessment or would just like to be more informed on the subject, you'll need some questions answered. You should know the type of terrorist groups that exist in the country that you are working in. Knowing this will help you know what attracts their attention. Ask questions such as:

Are they terrorists or criminals? If they are criminals using terrorist tactics you have a serious security problem that needs immediate attention.

Is there a pattern to terrorist activities in this country's past? The Tupamaro guerrillas (otherwise known as the MLN—*Movimiento de Libera-*

ción Nacional) of Uruguay were especially fond of robbing banks. Is there a preferred type of activity among your local terrorists?

Are they supported by a sympathetic third-party government, such as Libya or Syria? If they are a state-supported terrorist group they will have the finances to support themselves and more than likely not need to kidnap business personnel. These are groups to respect and take precautions against but not to the degree that you would for a group that needs money to support its activities. The latter type is far more likely to kidnap you for money. A state-supported group will kidnap you for political considerations, such as the release of their captured comrades. Both methods of operation need to be studied intensely. If the group in your area has kidnapped businessmen in the past, try to determine a pattern to their acts. Did they go into action after a lot of publicity about the company appeared in the media? Do they have a history of selecting one particular type of industry? If they always target oil companies and you work for an oil company, the threat level is obvious. Knowing your enemy requires a great deal of research. Assign someone in the company to examine everything the group has ever done, whether against business or government officials. If the terrorists have been successful, why have they been successful? How were they successful? Did their victim always keep a set pattern? Were the victims an easy mark? Why? Was their company always in the limelight? When examining these facts be brutally honest. What you are doing here is building your own "terrorist profile," just like the big boys do. You're finding out what type of person is going to come after you by looking at the details of those who were struck at before. If you, your company, or your company's personnel exhibit similar patterns, then it is definitely time to change the way things are done. Don't just look at the terrorists' successes. Examine their failures. If there have been attacks against corporations or businessmen that have failed, study them carefully and find out why they failed. Look for companies that have never had a man kidnapped. If they will let you, look at their security and see what they are doing. See if you can coordinate your security operation with theirs. In any battle, there is very definitely strength in numbers. As Benjamin Franklin once told his fellow patriots, "Gentlemen, if we don't all hang together, we will most certainly all hang separately."

DETERMINING THE RISK OF WORKING IN A COUNTRY

A country with a government opposing terrorism will be an easier country to do business in. The country must be willing and/or capable of dealing with the problem of terrorism.

A country in which anti-American sentiment is prevalent, where the government is not very stable, where there are problems between the government and the military, is a country that is ripe for terrorist problems.

Determining which of these two camps a given country falls into is a job for which you may have to rely on outside help. There are excellent information services available, among them private companies that produce volumes of data about the political situations in other countries. These companies do risk surveys and analyses and can provide you with detailed reports of the terrorist situation in whatever country you chose. These surveys can range from a simple outline of terrorist activity to a complete analysis of the political stability of a particular country. These companies and their services will be covered in detail later in the book.

When assessing the threat faced by companies versus your own personal threat, the first step is to look at terrorism on a global scale, since it is a global problem. A company should look at the same basic factors as an individual. Some of these factors are:

1. The number of social and/or political demonstrations in the country.
2. The number of terrorist incidents in the country and the subject of these acts. This is where risk analysis services come in handy.
3. The type of security forces within the country, and their effectiveness.

Your company should examine factors such as the country's crime rate and the overall attitude of the people towards political terror. For people in many places in the world terror is a fact of life, an accepted part of the political give-and-take.

In summary, the risk of terrorism is directly proportional to the amount of attention the company gives it. The more a company researches the problem, the less likely it will find itself a target of terrorism.

HELP FROM THE U.S. GOVERNMENT

Once you have analyzed your data, where do you go for help? The federal government is a good place to start. In the past few years, our government has begun to take a stand against terrorism. They have spent a great deal of time, money, and effort supplying the private community with information. They can supply you with information relating to the country you are doing business with, along with the security situation in that country, and how Americans there are generally treated. This information is obtainable from the Department of State or the Department of Commerce, and the United States embassy located in the country. Much of it is available free. Specific questions can probably be answered by the embassy security or political attaché. These people are there to help you, and generally do a good job. Since local laws can have an effect on your security program, especially when it comes to weapons possession and use, you should check with the embassy on the effects of these laws on your security plan. The following is extracted verbatim from Department of State Publication 8884, Department and Foreign Service Series 157 (January 1977):

> "American embassies and consulates will advise any American citizen or business representative who requests information on possible terrorist threats in foreign countries. The security officer or other designated officer at a diplomatic or consular post can provide the following information:
>
> - the nature, if any, of the general terrorist threat in a particular country;
> - whether private American citizens or companies have been the target of terrorist threats or attacks, in the recent past; specific areas of the cities or countryside that are considered dangerous for foreigners;
> - recommended host government contacts including police officials; local employment requirements for private security services;
> - methods and agencies available for security and background checks on local employees;
> - local laws and regulations concerning ownership, possession, and registration of weapons;
> - U.S. government policy on ransom and blackmail;
> - steps to take in case of a terrorist threat or act.
>
> In the case of a terrorist action against an American citizen or company, the embassy or consulate can:
>
> - facilitate communications with the home office and the family of the victim if normal channels are not adequate;
> - help establish useful liaison with local authorities;
> - provide information and suggest possible alternatives open to the family of the victim or his company. The U.S. government, however, cannot decide whether or not to accede to terrorist demands. Such a decision can be made only by the family or company of the victim, but it should be in consonance with local law. (U.S. policy, as publicly stated, is not to make any concessions to terrorists demands)".

If there is no U.S. embassy located near your place of work, you should contact the nearest consulate. Consulate staff monitor terrorist problems in their assigned countries, and can provide you with the same quality help as the embassy. Register yourself and your family with the consulate as soon as you set up housekeeping. Do this to get your name on any emergency evacuation plan. Ask for a briefing on the emergency plan, if one exists. If the country starts to have serious problems and an evacuation is required, there will not be time to sit down and study the plan. Be prepared. Plan ahead.

Any information from either the U.S. or your host government should be taken very seriously and action taken promptly. If they advise you a situation is dangerous, believe them. Do not hesitate. Heed whatever advice they give you. If you are told that a threat exists on your life by a member of the American embassy, attaché's office, or consular corps you should seriously consider taking yourself and your family out of the country. Embassy personnel will not make statements casually that affect your life. If they warn you, take the warning.

State Department statements are the result of information provided by reliable sources. More often than not, the embassy would rather not disclose where the information came from but don't discount the information's validity just because you don't know where they got it. There have been tragic incidents of companies and individuals who have not heeded warnings and paid dearly for their errors. Embassies are usually conservative in their estimates of problems. If you do not agree with their estimation of the situation, make sure your sources are better than theirs. To do anything else is just plain foolish.

LOCAL GOVERNMENT ASSISTANCE

If you feel sufficiently threatened, you should establish contact with government officials and the police of your host country at the highest possible level. Herein lies a problem: these government and police officials may be corrupt. If key officials or police officers are in league with a criminal or political terrorist group, it may make little difference or even be counterproductive to confer with them. Your best hope in this event is the development of your own personal intelligence network with contacts that can be trusted. If you have concern about the allegiances of these officials, have your personal intelligence network feel them out before you contact them directly.

In most cases, local government officials will be willing to assist you, although you must be cautious about the accuracy of their information. Some governments will "put on a happy face," and never admit to any danger, insisting instead that everything is fine and they have the situation under control. The more dependent the country is on the economic benefits provided by your company and/or your country, the more likely they are to distort their version of the local terrorist situation in order to keep both you and your business.

A good source of information on basic security practices relative to the local situation can be the local police department. Establish a good working relationship with a high-ranking police official. Make sure that whoever you select is a professional police officer and not a political appointee. The best method of establishing a relationship is to make him your friend. Be sure you trust him today, because tomorrow you may have to trust him with your life.

In many countries your company will receive special attention from the government. This gives you the opportunity to make a number of connections within that government. If you do, follow these guidelines:

1. Make inquiries through embassy officials, other corporate contacts, or security firms to ascertain which of the officials to be contacted are trustworthy, and are in a position to assist your firm.
2. Consider the capabilities of the country's security force against the company's needs. A bank might determine that the local police are equipped to assist it in its security problems, but a large facility, such as a refinery, might have to rely on the national security force. Quietly conducting a drill or some sort of low-key security practice is excellent procedure, but make sure to keep it from the terrorists' attentions. Otherwise it will become a dress rehearsal for your security operation, with your enemies as your audience.
3. Build on past goodwill. A local contact who has worked successfully with Americans in the past and is acquainted with their manner of doing business is more likely to respond to a corporation's problem.
4. Having established host government contacts, maintain them—even when there is no threat—to provide a link through which information may be exchanged should a terrorist threat become real or changes in government personnel and policy develop.

PSYCHOLOGICAL COSTS

Please keep in mind that terrorists or guerrillas do not have to stage an attack on the company to be successful. Their simple existence may cost the company dearly. If the living conditions are harsh and dangerous, morale will be affected, job performance will suffer, and profitable operations may become impossible. A Draconian security plan may cause unneeded problems, worsen morale, and inhibit profitable operations more than terrorists themselves.

If the security program is so rigid that the everyday routine of life is impossible, morale will suffer. Where danger, or perceived danger, is sufficiently severe (the level of severity differs with individuals) stress will take a major toll of executives, workers, and their families.

Any company involved in overseas operations should consult regularly with State Department officials to see what steps other firms operating in similar circumstances are taking to protect themselves.

In troubled areas, executives should make their presence known to the nearest American consulate. All employees leaving the U.S. for foreign operations, and their families, should receive at least a minimal briefing on security and the encouragement to practice it.

TERROR IN THE UNITED STATES

Since terrorism seems to be a spreading disease with no known cure, American companies are all asking the same question: "Will terrorism come to the U.S.? When will we have to worry about protecting ourselves here at home as well as abroad?" It's a difficult question to answer. But some answers can be gained by a quick glance at recent history. At 10:58 P.M. on November 7, 1983, an explosive device with a delayed fuse blew up in a hallway near the Senate chamber of the United States Capitol. The explosion opened a hole thirteen feet high in a thick inside wall, damaged congressional cloakrooms, tore the doors off minority leader Robert C. Byrd's office, damaged works of art, and hurled huge chunks of plaster through the air. This single bomb in the U.S. Senate caused an estimated quarter of a million dollars' worth of damage. There were no injuries, but that was only through sheer luck. The Senate had planned a late night session on November 7, but Senate Majority Leader Howard Baker decided that afternoon to cancel it. Had the Senate still been in session that hall would have been filled with senators, aides, and other personnel. The result would have been catastrophic.

A group calling themselves the Armed Resistance Unit took credit for the blast, saying that the bomb was in retaliation for the U.S. actions in Lebanon and Grenada. This same group took credit for a bombing at the Army War College at Fort McNair in Washington, D.C. in 1983. According to FBI director William Webster, ten bombs exploded in federal buildings during a recent 18-month period. It seems that corporations should not be concerned over whether terrorism will eventually rock the U.S., but wonder instead how severe the onslaught will be once it starts. There is an overall public perception that there is very little terrorist activity within the U.S. today. Brian Jenkins, one of the nation's foremost authorities on terrorism, feels that this perception is due to the fact that it takes a spectacular act of violence to get the attention of the American media. We are a violent society, with over 20,000 homicide deaths a year. Unfortunately, it requires an especially depraved act of wanton violence, such as the 1985 murder of wheelchair-bound American tourist Leon Klinghoffer aboard the hijacked Italian cruise ship *Achille Lauro* to capture the attention of the American people.

3

How Terrorists Choose Their Targets

THE TERRORIST SELECTION PROCESS

By now you have completed your threat assessment and decided that the country where you work has a definite terrorist problem. Worse yet, you fit the terrorist profile. There's no need to panic. Just because you fit the profile does not necessarily make you a target. It may put you on a list of potential targets, however, but to become an actual victim you will have to become the final candidate in what is known as the "terrorist selection process." Terrorists are very methodical and very fussy in selecting their targets. They cannot afford a failure.

So relax, at least relax a little. Fitting the terrorist profile only means that you may become part of their selection process. And what does that mean?

The selection process is a series of steps taken by terrorists to develop a list of potential targets. By using a well-organized selection process terrorists will outline a group of people as possible candidates for kidnappings. More often than not, terrorists do not get emotionally involved with their victims. They usually have nothing personally against the individuals they chose. They are more interested in what the potential victims represent and what kidnapping them will get from the companies their victims work for. What they want is usually money. Sometimes "confessions" or admissions of wrongdoing are thrown into the deal for local political considerations.

A point in your favor is the fact that in any given country at any given moment there is an abundance of people other than you that fit the terrorists' profile. As far as the terrorists are concerned, any one of those people will accomplish the same goal. For example, if you are the manager of an American oil company and the local terrorists want to strike at an oil company, chances are there are plenty of other American oil company executives to choose from, to analyze, evaluate, and select as a target. It's rather like being a candidate for a job you didn't know you were being evaluated for. You and your "resumé" are looked over by a "personnel selection board" made up of terrorists. They carefully run you and your fellow executives through a selection process and the hapless "winner" gets kidnapped.

It's a topsy-turvy job hunt you don't want to win, for the simple fact that in winning, you lose. Therefore, knowing how the selection process works is a powerful asset for your personal protection program.

THE TERRORIST SELECTION PROCESS: PHASES I AND II

The terrorist selection process is divided into four phases. During Phase One the terrorists select a group of potential targets that fit the terrorist profile mentioned in Chapter 2. In any foreign country there are a number of people who will fit the terrorist profile and will be picked in this first phase of the selection process. Once a group of targets has been identified, the terrorists move on to Phase Two, whittling down the list of potential candidates through a process of elimination. This is done by meticulously gathering information about the day-to-day routines and habits of each potential target. It is critical that you understand how this information-gathering process works. If you understand how terrorists select their final target, you can develop a personal security program which includes effective countermeasures. During this phase of the program, terrorists gather information by conducting long, detailed surveillance of each of the potential targets. If you fit the terrorist profile and have been selected as a possible target, be assured you will now be kept under surveillance. The results of the surveillance will be volumes of data on you and your fellow target candidates. The most important information produced from this surveillance data is the daily time schedule of each potential target. Terrorists will carefully monitor your daily routine, accurately outlining your every move during your daily routine. They'll document what time you leave for work, what type of car you drive, what route you drive to work, where you go for lunch, what time you go for lunch, maybe even what you had for lunch, where you go for leisure-time activity. They'll carefully examine your routine, looking for a pattern.

Do you always leave for work at the same time? Do you always play tennis at the same time? They'll know your routine better than your own family, maybe even better than you.

Sometimes terrorists tip their hand during this surveillance process. Suspicious cars filled with mysterious people following you around can draw attention. Squads of Libyans shadowing American diplomats recently helped draw attention to that country's terrorist intentions, and perhaps helped alert American embassies to the potential danger of attack.

We'll talk about how to tell if you're being followed and what to do if you are in a moment.

Once it's been established you are personally vulnerable to kidnapping, that is, that it's operationally possible to nab you, the terrorists will continue their research to determine if you fit their political/financial needs. This evaluation of your worth as a target constitutes Phase Two of the program. If the terrorists need money, they'll look for evidence that either you have substantial personal assets or your company is willing to pay a stiff ransom to gain your release.

In the last chapter we learned that terrorists look for people who show signs of wealth. However, money may not be their only motive. Kidnapping someone who is doing harm to their country makes for good propaganda. You must evaluate your company's image abroad, as well as your own. Have there been articles in the local newspapers and magazines accusing your company of polluting the country's environment? Are they true? Has anything that you or your company done created a negative image with the public? There are times when everyday company discipline may be interpreted locally as being excessively strict with the workers. Don't overreact to negative publicity; an unflattering article about you or your company in the local paper does not mean you're an automatic target for terrorism. But more than likely, if you're living in a terrorist environment, adverse publicity means that your name has just been moved up the list. And that's a hit parade in which you do not want to be number one.

PHASE III: OF THE SELECTION PROCESS

Once they've gathered all the information they can about their potential targets, the terrorists will proceed to Phase Three. During this phase they make a very careful risk analysis of each potential target. After careful analysis of their surveillance data, they'll finally decide which kidnap candidate will be the easiest to nab; which will offer the smallest risk. For instance, let's assume they have decided on an oil company executive as their target. There are two oil executives on their potential hit list. From data collected in Phase Two of the selection process, they notice that one of the two oil execs keeps a very regimented routine, leaving the house every morning for work at the same time, plus or minus five minutes, walking out the front door and climbing into a chauffeured car. The chauffeur takes the same route to work every morning. Every morning at the same time, plus or minus five minutes, the driver will be at the same corner stopping at the same stop sign. Our target executive has lunch at the same restaurant at the same time every day, arriving home at the same time every night, going to the health club every Wednesday night, playing golf every weekend. With this type of individual, it's possible to merely glance at your watch and say "it's Thursday, and it's 8:15 A.M. He must just be leaving the house." He's easy to find; it's easy to know where he'll be and when he'll be there. In other words, he is a pigeon ripe for the plucking.

The other oil company executive on the selection list is unpredictable. He varies his routes to and from work, and keeps an unpredictable time schedule. It's hard to fix a time when he'll be at work or at home. It's hard to predict where he'll be and when he'll be there.

The choice for the terrorist is an obvious one. By the simple rationale of predictability, they'll select the first man as the target. Terrorists operate just as any good company would if it were about to enter into a business venture. As a company wants to be assured of success, so do terrorists. They conduct a feasibility study. They select their final target, a target of opportunity against which the terrorists feel they'll enjoy the highest potential for success. The tremendous success rate of terrorists against businessmen should come as no surprise. They simply do their homework. What may seem like a "daring daylight attack," as the press often puts it, was in fact not a daring attack at all, merely the result of good research, competent planning, and a complacent and therefore vulnerable target. What the public and the media witnessed was a well-planned and -executed attack on a well-selected target, a target that was the result of meticulous research. And research obviously pays. Thanks to research, the terrorists can go into an operation in which they have chosen the target, along with the time and method of attack, with a high probability of success. That is the basic reason why terror operations are so often successful.

PHASE IV OF THE SELECTION PROCESS

Once the terrorists have selected their final target, they still continue with their planning. They know that the key to success is that the victims must be totally unaware they are targets, in order to be taken completely by surprise.

To insure their actions will be a total surprise, the terrorist will go into a fourth phase, which includes additional surveillance of their victim. The target's movements will again be analyzed and a pattern of habits reestablished. Sometimes they'll actually take photographs of the target and actually practice making the attack. It is during this fourth phase of surveillance that terrorists will select a time and place for the kidnapping. The time and place will be precisely at that point in the executive's daily routine that insures the highest possible chance of success. For our oil company executive who keeps a carefully regimented routine, the time of the kidnapping may be in the morning after he leaves for work. The place may be in his car as it stops for traffic at an intersection near his house. Surveillance shows he'll be in his car, at that intersection, at that precise time. Look at it as a football game. The terrorists will find the manager's weak point and hit him there.

If there is anything you can know for certain in a terrorist kidnapping, it is that if you are the selected target, the terrorists will pick a location where they are 100 percent sure you will be, at a time they know with 100 percent certainty you will be there. If you think you might be on a terrorist selection list, examine your daily routine carefully and make an honest assessment of your habits. Write down your daily routine carefully noting the times and the places you visit.

Then take that data and scramble your schedule as much as possible. If you leave for work every morning at 7:15 A.M., and always make a left out of your driveway, make a right the next time. Leave early. Leave late. Use a different car from time to time. Stay home someday and see if that blue Chevy parked out front—you know, the one that always seems to pull out of its parking place right after you pass by—see if that car doesn't stay put for most of the day.

Be unpredictable. If you are at an intersection every morning at 7:22 A.M. you can almost bet that the kidnapping will take place at that intersection. This is one reason why so many kidnappings take place while the victim is in a vehicle. The other reason is that an individual is surprisingly vulnerable while traveling in a vehicle. Journeying to and from work by car at roughly the same time every day quickly becomes an inalterable, easily predictable personal routine, making it easy for terrorists to predict where you will be at a given moment. (This subject will be covered in greater detail later in the book). But all is not lost—practicing good security, whether in the office or your vehicle, can make you less of a target, making you just a little bit harder to kidnap than the next guy.

Sometimes good security can be something that may appear trivial and sometimes it can be an outright accident. There are recorded cases in which terrorists did not attack a particular car because they could not tell if an additional antenna on the car was a device that would activate a vehicle defensive system, or was connected to an emergency-band radio. An innocuous thing like an antenna stopped the terrorists, forcing them to divert to another target. Terrorists are very cautious: anything that looks suspicious can be enough to make them look someplace else. Terrorists are usually in no rush, at least successful terrorists who want to live and therefore will wait for their chance to get at the easiest target.

Since terrorists almost instinctively seek out weak security arrangements, your own security has to be at least equal to or greater than the security of other executives working in the country. Terrorists will go to extreme lengths to evaluate your security, and having make-believe security will not fool them. Some executives have gone as far as adding bodyguards who follow in a car behind the executive's as a security measure. If they are not well-trained, competent bodyguards, and look it, the terrorists will know. They'll analyze the added risk of kidnapping a person protected by bodyguards, examining the skill of the bodyguards, what type of weapons they carry, and whether they feel the bodyguards will offer any real deterrent to their attack.

Former Italian president Aldo Moro was guarded by bodyguards when he was kidnapped in his vehicle and eventually murdered. Five of his bodyguards died in the attack. He felt he had adequate protection, but so did the other possible targets on the terrorists' list. The other people on the selection list had what the terrorists considered to be good bodyguard teams as well. What turned the terrorists away from the others and made them select Moro as their final target was that the other candidates for kidnapping were driving *armored* cars. The Red Brigade terrorists felt the risk of attacking an armored vehicle protected by bodyguards was too great, so they decided to go with Moro, the easier target. The Red Brigade hit team felt Moro's bodyguards would offer no real obstacles. Unfortunately for Mr. Moro and his bodyguards, the terrorists were right. It pays to do your homework.

DON'T LOOK NOW BUT . . .

If you are on a terrorist list you will definitely be put under surveillance. This is not casual surveillance, it is intense. If you are selected as the final target, you may be under surveillance for weeks. Captured documents indicate that if terrorists decide to kidnap you while in a vehicle, they'll often shadow your trip to and from work for quite some time. In one case terrorists followed a car over the entire route to and from work over a space of five weeks. Candid photos of you during your daily routine are also part of their surveillance. In another case, police uncovered a photo album documenting the kidnapping candidate's every move over an entire week. This group also had photos of those people that came into close contact with the victim and his family during that time. The terrorists had planned to plant their own people along the day-to-day routine of their target to aid in the attack.

A classic example of the problem of predictability is the case of a Turkish diplomat who was assassinated while driving home from work. Before the attack, the killer was seen sitting in a doughnut shop reading a newspaper. A video camera located in the doughnut shop observed him checking his watch periodically. At the precise time the Turkish diplomat was to drive by, the assassin got up and took his position alongside the road. As the diplomat slowed down to turn a corner, the assassin fired through the car window and killed the diplomat.

What killed the diplomat wasn't the assassin's bullet; it was his predictability. The assassin knew exactly what time his target was to drive by and knew that he would have to slow down for the corner. That brief moment of vulnerability was all the assassin needed.

HOW TERRORISTS ARE TRAINED

Whether a terrorist group attacks a structure—such as a building or power line—or attempts a kidnapping generally depends on the size of the group that will actually perform the attack. Terror against a target such as a building can be carried out by an organization of any size—a single person can easily build and plant a bomb in a building. A kidnapping will be carried out by a larger group for the simple fact that a successful kidnapping requires a good deal of people. Ideally, the group separates into teams responsible for different parts of the attack. One team may be responsible for the actual grabbing of the victim, while another team is responsible for setting up the scenario. There is one team of terrorists to handle negotiations, and another to deal with communications.

It is obvious that successful terrorists are well organized. They are also well trained. The training comes from a variety of countries that support worldwide terrorism. Training terrorists is little different than any other paramilitary training process. First, the students need a training manual. The preferred manual by far is Marighella's *Mini-Manual of the Urban Guerrilla.* This training manual was mentioned in Chapter 2. The significance of Marighella's manual should not be overlooked. Back in the late 60's Marighella predicted that the way to defeat capitalism is to take the fight to the city streets. He studied the works of various revolutionaries and adapted these techniques to an urban environment. Any executive working in a terrorist-infested environment must understand Marighella's work. Reading it is like studying the competition's game plan.

Marighella's action models can give you a better understanding of how terrorists organize their attacks. His terrorist technique is aggressive, using hit-and-run tactics. Successful terrorist actions require several initial advantages over the target, including surprise, superior knowledge of the area where the action will occur, greater mobility and speed than police or security forces, complete information on the target, and total command and control of the action while it is being carried out.

The Marighella manual stresses the need for the kidnapping to be a total surprise. Marighella describes the technique of surprise as:

> a. we know the situation of the enemy we are going to attack, usually by means of precise information and meticulous observation, while the enemy does not know he is going to be attacked and knows nothing about the attacker; [NOTE: This is the reason for the meticulous surveillance on the subject they plan to kidnap.]
> b. we know the force of the enemy that is going to be attacked and the enemy knows nothing about our force;
> c. attacking by surprise, we save and conserve our forces, while the enemy is unable to do the same and is left at the mercy of events;
> d. we determine the hour and the place of the attack, fix its duration, and establish its objective. The enemy remains ignorant of all this.

Keep in mind that throughout Marighella's outline, you are the enemy. He is saying they have to have precise information about "you," and rely on the fact that "you" don't know anything about them, and that "you" are at their mercy. Closely related to surprise is another essential: information. Terrorists cannot afford to take spontaneous, unplanned actions; the terrorist group is nearly always small and weak and can be wiped out by a larger, prepared force. Planning facilitates surprise, which is the key ingredient for a successful terrorist or guerrilla action. Planning is impossible without accurate information about the target. For the terrorist, information must be collected on the target's movements, habits, protection, everything related to his vulnerability. Marighella supplies techniques for gathering intelligence:

> The urban guerrilla, living in the midst of the people and moving about among them, must be attentive to all types of conversations and human relations, learning how to disguise his interest with great skill and judgment.In places where people work, study, and live, it is easy to collect all kinds of information on payments, business, plans of all types, points of view, opinions, people's state of mind, trips, interiors of buildings, offices and rooms, operation centers, etc.

This is a good reason to be careful about who you hire as domestic help, or as cleaning people in the office, or any other activity that requires people working on a short-term basis in your environment.

Marighella continues:

> Careful reading of the press with particular attention to the organs of mass communication, the investigation of accumulated data, the transmission of news and everything of note, a persistence in being informed and in informing

others, all this makes up the intricate and immensely complicated question of information which gives the urban guerrilla a decisive advantage.

The *Mini-Manual* gives instruction on how to carry out an action. Terrorists must double-check their information, and if they have not obtained information elsewhere, they must perform detailed reconnaissance of both the victim and the terrain. They must also study and time routes; vehicles, if used, must be carefully chosen. Personnel selection also requires great care, as does the selection of weapons. Rehearsals of the action coupled with study are necessary for a successful action.

HOW TERRORISTS OPERATE

Carlos Marighella's *Mini-Manual for the Urban Guerrilla* provides a number of so-called "action models" that terrorists can use against governments and businesses. Here are some examples:

Execution: Execution is the killing of a North American spy, or an agent of the dictatorship, of a police torturer, of a fascist personality in the government involved in crimes and persecutions against patriots, of a stool pigeon, informer, police agent, or police provocateur. (NOTE: This is the reason that most terrorist groups do not assassinate executives. Business executives don't fit into any of the above categories, or never should and never should have anyone think they do.)

Kidnapping: Kidnapping is capturing and holding in a secret spot a police agent, a North American spy, a political personality, or a notorious and dangerous enemy of the revolutionary movement.

Kidnapping is used to exchange or liberate imprisoned revolutionary comrades, or to force suspension of torture in the jail cells of the military dictatorship.

The kidnapping of personalities who are known artists, sports figures, or are outstanding in some other field, but who have evidenced no political interest, can be a useful form of propaganda for the revolutionary and patriotic principles of the urban guerrilla provided it occurs under special circumstances, and the kidnapping is handled so that the public sympathizes with it and accepts it.

The kidnapping of North American residents constitutes a form of protest against the penetration and domination of United States imperialism in our country. (NOTE: This is why they kidnap executives.)

Assaults: Assault is the armed attack which is made to expropriate funds, liberate prisoners, capture explosives, machine guns, and other types of arms and ammunition.

Assaults can take place in broad daylight or at night.

Daytime assaults are made when the objective cannot be achieved at any other hour, as, for example, the transport of money by the banks, which is not done at night.

Night assault is usually the most advantageous to the urban guerrilla. The ideal is for all assaults to take place at night when conditions for a surprise attack are most favorable and the darkness facilitates flight and hides the identity of the participants. The urban guerrilla must prepare himself, nevertheless, to act under all conditions, daytime as well as night time.

Raids and penetrations: Raids and penetrations are quick attacks on establishments located in neighborhoods or even in the center of the city, such as small military units, commissaries, or hospitals, to cause trouble, seize arms, punish and terrorize the enemy, take reprisal, or rescue wounded prisoners, or those hospitalized under police vigilance.

Raids and penetrations are also made on garages and depots to destroy vehicles and damage installations, especially if they are North American firms and property.

When they are carried out in certain houses, offices, archives, or public offices, their purpose is to capture or search for secret papers and documents with which to denounce involvements, compromises, and the corruption of men in government, their dirty deals and criminal transactions with the North Americans.

Occupations: Occupations are a type of attack carried out when the urban guerrilla stations himself in specific establishments and locations for a temporary resistance against the enemy or for some propaganda purpose.

The occupation of factories and schools during strikes or at other times is a method of protest or of distracting the enemy's attention.

Ambush: Ambushes are attacks typified by surprise when the enemy is trapped across a road or when he makes a police net surrounding a house or an estate. A false message can bring the enemy to the spot where he falls into the trap.

The principal object of the ambush tactic is to capture enemy arms and punish him with death.

The urban guerrilla sniper is the kind of fighter especially suited for ambush because he can hide easily in the irregularities of the terrain, on the roofs and the tops of buildings under construction. From windows and dark places, he can take careful aim at his chosen target.

Ambush has devastating effects on the enemy, leaving him unnerved, insecure, and fearful.

Sabotage: Sabotage is a highly destructive type of attack using very few persons and sometimes requiring only one to accomplish the desired result. When the urban guerrilla uses sabotage the first phase is isolated sabotage. Then comes the phase of dispersed and generalized sabotage, carried out by the people.

Well-executed sabotage demands study, planning, and careful execution. A characteristic form of sabotage is explosion using dynamite, fire, and the placing of mines.

> North American firms and properties in the country, for their part, must become such frequent targets of sabotage that the volume of actions directed against them surpasses the total of all other actions against vital enemy points.

Terrorism: Terrorism is an action, usually involving the placement of a bomb or fire explosion of great destructive power, which is capable of effecting irreparable loss against the enemy.

Although terrorism generally involves an explosion, there are cases in which it may also be carried out by execution and the systematic burning of installations, properties, and North American depots, plantations, etc. It is essential to point out the importance of fires and the construction of incendiary bombs such as gasoline bombs in the technique of revolutionary terrorism. *Another thing is the importance of the material the urban guerrilla can persuade the People to expropriate in moments of hunger and scarcity resulting from the greed of the big commercial interests.*

Terrorism is an arm the revolutionary can never relinquish.

The war of nerves: The war of nerves or psychological war is an aggressive technique based on the direct or indirect use of mass means of communication and news transmitted orally in order to demoralize the government.

The object of the war of nerves is to misinform, spreading lies among the authorities, in which everyone can participate, thus creating an air of nervousness, discredit, insecurity, uncertainty, and concern on the part of the government.

COUNTERSURVEILLANCE

If you have been selected as a potential target you still have an excellent chance of not being selected as the final target. One of the ways you can do this is by careful examination of the Marighella *Mini-Manual*. From examining the *Mini-Manual* it becomes apparent that surveillance is the key to the success of any terrorist kidnapping. A good way to make yourself less likely to be chosen as the final target is to detect the surveillance. This is not as hard as one may imagine. It is simply a matter of developing an awareness of what's going on around you. The British ambassador to Uruguay, Sir Geoffrey Jackson, was kidnapped from his car and spent nine months in a "People's Prison." Well before the actual kidnapping he became aware of changes in his surroundings. In a park across the street from his house, he noticed there always seemed to be a young couple with a baby having a picnic. He also noticed from time to time that even though there were different couples in the park having their picnic, they all followed the same pattern. Rather than focus their attentions on their baby, the way most young couples would, these couples seemed to watch his comings and goings with great interest.

On other occasions, while driving to work, he noticed a motorscooter persistently cutting in front of his car. The scooter always carried a young girl. A procession of different people seemed to be driving the same scooter. Sir Geoffrey noticed this because he noted the scooter's registration number. What Sir Geoffrey was watching was rehearsals for his own kidnapping. The terrorists were actually practicing right before his eyes. The scooter would routinely cut him off. A car or a van would conduct "practice ambushes" to decide the best way to block his car. When he got to the embassy, Sir Geoffrey often noticed a pair of students hugging and kissing, but oddly more interested in him than in each other. All of these are clear signs that Sir Geoffrey was under surveillance. Unfortunately for our British diplomat, he observed these suspicious activities, but did nothing about them. The signs went unheeded, with his kidnapping the result.

SURPRISE AND THE EARLY WARNING SYSTEM

The necessity of surprise and the gathering of reliable information is stressed throughout Marighella's manual. Terrorists know that these two elements are the key ingredients in a successful kidnapping. However, if the kidnapping is to be successful the most important, dangerous, and delicate part of the kidnapping is the actual taking of the victim. For a kidnapping to be a success the victim must be kept completely unaware of what's about to happen. If the terrorists feel that they cannot surprise the victim, they'll not select him as their final target. Therefore, examining it from the potential victim's point of view, the logical thing to do is to reduce the element of surprise. You can protect against unpleasant surprises by analyzing the tactics of surprise.

Surprise requires planning. To surprise you, the terrorists first need to keep you under constant surveillance. Therefore, you develop a surveillance awareness plan as a countermeasure against surprise. This gives you a tremendous edge over other potential targets who are not as aware of their surroundings.

To develop an early warning system, become more sensitive to your surroundings. The best place to start building your early warning system is near your home and office. There is an amazing yet reliably true statistic that indicates that the areas around your home and office are the most dangerous locations in your daily routine. Most kidnappings occur near the home or the office, which only makes sense since those two places are where you spend most of your time. Terrorists pick these areas to conduct kidnappings because it is difficult for you to be unpredictable near your home or office. You usually have a fifty-fifty choice in the morning. Once you get into your car and leave your driveway, you can go to the left or right. Your choice of alternative routes in the immediate vicinity of your home is very limited. The farther you get away from the home or the office the more alternatives you have toward altering your routes and other parts of your routine. So the terrorists' choice of and ideal place to strike at you are really rather limited, if they want to play their game conservatively.

Don't be misled by terrorists' propensity for striking at you at home or on the way to the office. It can happen any time you are in public. Swedish prime minister Olof Palme was strolling home with his wife after dismissing his personal bodyguards following a night out at the movies. Evidently Mr. Palme felt quite safe on that peaceful Stockholm street—that is, until a gunman mowed him down before his horrified wife.

The next element of the kidnapping equation is time. More than likely you did not get to your present place in your career by showing up late for work. By following the same route to work every day you will pick the location of your kidnapping yourself. By being punctual, you will also select the time. Both these factors can work to your advantage, however. The key is to be unpredictable. If you are unpredictable, it will be difficult for your terrorists to remain inconspicious during the surveillance. Develop your early warning system by keeping a close watch for abnormal activity near the home and office. At the same time, develop some abnormal activities of your own. Take various routes to work in the morning, likewise on the trip home. Notice if certain vehicles seem to be following you. You might even try to lose them, in an inobtrusive way that observes local traffic laws, of course. This may sound like a lot of work, but it's really not much effort.

Most people have this type of early warning system and don't realize it. To develop your warning awareness, when you leave your home in the morning, make some notes, either on paper or in your mind, on the activities you observe around your home. A pocket memo tape recorder is useful for this type of note taking. Record the types of vehicles you see in your neighborhood at various hours of the day, along with children going to school and people you happen to see on the street regularly, especially new faces. Within a few blocks of your home you usually see the same people, cars, etc., in roughly the same places. In fact, most people can tell whether they are late or early by who and what they see on the street.

Paint a mental image in your mind of what you should see, and you'll be far more likely to spot the things you should not see, when and if they appear.

If a new element is introduced into your mental picture, a signal goes off in your mind. You'll ask yourself, "What is that doing there?" "Who is that person?" Is there a strange car parked across the street? Are there unfamiliar people in the area? A vehicle or group of people that just does not belong in the neighborhood?

Most VIP's live in affluent areas—a fact which usually dictates the general type of vehicle and people that will be present in the area. Anything out of the ordinary will stand out plainly.

What you are doing with this process is quite simple: You are subconsciously comparing what you are seeing to what should be there, based on careful observation over a fairly long period of time. When an object, person, or unfamiliar scenario comes into the picture that does not equal what you should see, a question will be raised—what is that doing in this area? Frequent practice with this technique will develop your ability to spot routine, often ignored aspects of the day-to-day landscape that can easily conceal hidden dangers which could threaten your life. If Sir Geoffrey had thought it odd that a strangely distracted series of picnicking couples were always across the street from his residence, and had done something about it, he might not have had to undergo the trauma of kidnap and nine months of captivity.

You must raise your level of awareness to a point where:

a. Strange vehicles parked near your residence or place of employment are noticed and promptly reported to the authorities. *This must be done immediately*. It may be the first time you have seen the vehicle but it may not be the first time it has been there. You don't know what level of planning the terrorists may be in. Maybe it's only the beginning, but perhaps their planning is in the final stages. That gives you little time to act.

b. People standing, walking, or sitting in cars near the residence or place of employment must be noticed, especially people loitering.

c. Someone who always seems to be around you. An alert individual can realize when he is being followed.

It's important to keep track of any unusual sightings. A small, hand-held tape recorder makes it easy for you to record any unusual activity. As a busy executive, you probably make use of one already to record ideas, memos, and notes. From the tape, a log can be made describing what you see.

(An additional payoff is that to an observant terrorist, it may look as if you are speaking into a portable transceiver, maybe even talking to the police! Such appearances have been known to make terrorists uneasy.)

When taping your observations of suspected terrorist vehicles, make sure you get the following information:

1. Make
2. Model
3. Year
4. Color
5. License number
6. General condition of the vehicle
7. Number of people in the vehicle

If possible, get a description of the people involved, such as:

1. Male/female
2. Distinguishing marks, beards, mustaches, scars, etc.
3. Size
4. Features such as hair style and color are often unreliable, because they are the results of wigs, dyes, etc. But make note of them anyway. Too much information is much better than too little.
5. Distinctive features or habits: a limp, a nervous habit such as frequent straightening of a tie, pushing glasses up onto the nose, etc.

KNOWING IF YOU ARE A TARGET

Spotting those assigned to shadow you is your surest indication that you are a potential target. Letting those shadowing you know you've spotted them is the surest defense against them attacking you. Terrorists' best defense is their anonymity. When this protection is removed, they become naked and extremely vulnerable. If your personal security program is well planned and you adhere strictly to it, it may simply mean that terrorists will decide your security is good and that an attack on you might be very difficult, not to mention dangerous, and therefore not worth the effort. Any indication of surveillance, no matter how seemingly trivial, should always be taken seriously. When surveillance is detected, report it immediately to your security department and the U.S. Embassy. If you do not have a security department, report it to the local police (if they can be trusted) and to the U.S. embassy. Provide whomever you talk to with as much information about the surveillance as possible. Increase your security and eliminate all unnecessary trips away from your home or office. If possible, take a vacation out of the country. Watch for any continued surveillance; have your friends and family help you.

Detecting surveillance may seem a difficult process but this is not generally the case. Quite the contrary, good surveillance is a difficult task. Good professional surveillance teams are generally limited to those employed by governments, elite law enforcement and intelligence organizations, and, unfortunately, a select group of very successful terrorist organizations. However, even these surveillance teams make mistakes or perform actions that make them detectable by their target.

Detecting surveillance is the simple matter of knowing who you should be looking for, keeping your eyes open, and, above all, not believing in coincidence. For instance, if you see a red Ford with a dented right fender parked outside your house in the morning, and you have never seen this vehicle before, it may be a coincidence. There could be any number of explanations why it is there. If you see the same vehicle the next day on your way home, it could possibly mean surveillance. If it follows you any time within the following few days, it is almost assuredly a surveillance. And if that same red Ford is always parked across from your house, call in your security network.

Don't become fixated on any one vehicle, however. Clever terrorists employ a small fleet of nondescript vehicles. The Tupamaros that kidnapped and executed American police advisor Daniel A. Mitrone routinely used a makeshift fleet of vehicles "commandeered" from Uruguayan civilians, giving them a constantly changing selection of vehicles and making those urban guerrillas very hard to track.

Most surveillance efforts will start at your house and follow you to your final destination. This is why it is so important to know what does and does not belong around your house. Surveillance will generally cover your entire day's routine up to your final arrival at home. It may take a variety of forms but will generally include surveillance by more than one automobile, van, truck, or other vehicle. Practice all of the following guidelines carefully. They are simple and require very little concentration.

Make notes as to description of both people and vehicles, as well as the time and place.

Do not do anything to let the surveillants know they have been detected. You do not want to confront a couple of terrorists and ask them why they are watching you. You may not like their response.

If they continue with the surveillance, notify your security department and local U.S. embassy. Give them all the information you have and let them make the decisions.

The majority of all surveillances are conducted by terrorists in a car. Keep an eye out for anyone, pedestrians or people in parked vehicles, who are watching as you enter or leave your car. Remember, your car is the focal point of the pending attack. Anyone following you during the first few blocks of your home or office could be conducting a surveillance. Watch your mirrors.

As you drive, be aware of any car that pulls out of a parking place, side street, or driveway and appears to be following you. Surveillants will not always be following from the rear. They can also be watching you from a vehicle in front of you. If someone is observing you from the front their mirrors may be adjusted so the passenger can see you. If you look into the rear- or side-view mirrors of the car in front of you and see the face of the passenger looking back at you, you should be suspicious. Note the car, its license plate, and get a description of the occupants.

Learn to memorize license plates, as well as makes and models of cars at a glance. In a foreign country you will encounter many types of cars that we don't see much in the U.S. So learn your car types. This is a vital way of determining if a pattern is developing.

Those keeping you under surveillance will generally not follow you right to your destination. If they know your final destination and route, they'll probably break off the surveillance just before you get where you're going. At this stage of the game they do not want to risk a personal confrontation with you. Not yet.

It cannot be stressed strongly enough that if you feel you are under surveillance, immediately discuss the matter with your security staff. If you have no security staff, and think someone is following you, drive in a random fashion some time when you think you are being followed. Don't play James Bond with the terrorist by making illegal U-turns, plowing through red lights, or driving through the local farmers' market at 95 mph, but drive a random pattern of streets that would be extremely unlikely for anyone else to drive. If the suspect car is still behind you after, say, a half-hour's aimless wandering, it's pretty likely you are under surveillance. Should this be the case, *don't panic*. Drive home calmly and immediately alert your personal security network that you have become a terrorist target.

All this sounds like you should become paranoid. Don't. Excessive paranoia is no good for you. A little bit of anxiety is good for you, however, especially if you live in a terrorist environment. Anxiety will keep you alive. Excessive paranoia will keep you jittery and prone to do something stupid.

PERSONAL SECURITY: AN INTRODUCTION

Now that you have begun your threat assessment (a process that never really ends), and have decided that you are a strong candidate for a terrorist's "potential target" list, your next step is the creation of a personal security program. While no one likes to think about it, personal security is something you always need. No matter what country you live in, everyone needs some form of personal security, not only to guard against terrorism but to protect against all forms of violence. There are certain sections of towns in the U.S. where you really should not go after dark. So you use common sense and stay away from those areas. When you do this, you are employing a simple form of personal security. Locking our doors at night is another form of personal security.

So personal security is a big part of your daily routine. But personal security takes on a whole new meaning when you work overseas. If you are working for an American company overseas, you face a unique sort of personal security problem that most Americans at home in the U.S. are unfamiliar with. When you work in a foreign land as a representative of an American corporation, you must understand that along with representing your company you represent the United States of America. Therein lies your problem. Given the tremendous increase in terrorism around the world, the simple fact of your "American-ism" makes you a potential target. If you fit the "terrorist profile" and are unfortunate enough to be selected as one of the finalists, your only defense is to develop a personal security program, a program that meets your needs, a program you will follow religiously. No personal security program will work if you don't believe in it. Good personal security begins with a good attitude.

THE NEED FOR PERSONAL SECURITY

Due to the high standard of work performed by police departments throughout the U.S., Americans tend to think that our country can and should protect us anywhere we go at any time. This is not the case. For proof, consider the recent massacres at European airports, along with tourist hijackings, and bombings at public facilities around the world. If history has taught us anything over the last few years it is that when we are working or travelling in another country there is a limit on the protection the U.S. can offer its citizens. America, or any country for that matter, cannot guarantee protection for its citizens wherever they go. Sadly, there is a clear and present danger for any American who leaves the U.S. to work in another country. That doesn't mean that every country you visit or work in is a place that will require you to have an organized personal security program. There are many places in the world that are safer than any big city in America. However, as a manager for an American corporation and as a family man you must be aware of the dangers involved in working in certain parts of the world.

Most Americans do very little to protect themselves once they leave American shores. We are a trusting people, and sometimes feel there is some mystical power watching over us to keep us from harm. This is sheer nonsense. Working overseas requires you to organize your safety and the safety of your family. Your company can supply you with state-of-the-art hardware; alarms, motion detectors, bomb detection equipment, armored cars, and communications. They can hire security consultants to supply training intended to protect you and your family. While these consultants can advise you, you are the person to make their advice successful. You have to take charge, and guide your own destiny and that of your family. Consultants and the company's security department can supply you with information, but you are the one that has to implement whatever the consultants ask you to do. If you ignore their recommendations, you are courting disaster. Most companies spend a great deal of time and money offering their overseas employees security assistance. Security departments in multinational companies are staffed by competent security professionals who are willing to give you 110% performance at all times, supplying you with all the help necessary to get the job done. Their job is to make your stay in a foreign environment as safe and productive as possible. This can be accomplished much more easily if you have a basic knowledge of personal security. This knowledge will both help you and allow you to communicate your needs with security personnel in a much more professional manner. Of course, all this assumes that you have a security department or consultants at your disposal. If you don't, then a practical understanding of personal security is of infinitely greater importance to you.

ASSESSING YOUR PERSONAL SECURITY NEEDS IN THE COUNTRY

To develop a workable personal security program, you must begin by learning as much as possible about the country in which you'll be working. If you are preparing to move overseas, there are services available that will help you and members of your family learn about and adapt to your new environment. These companies can supply you with information about the customs and cultures of the country, as well as provide you with a security briefing on the country. This is important. Moving into a new country you may encounter situations that can make your first few months a nightmare.

You and your family should learn the culture, and customs of your host country. Something that may be an everyday experience in the U.S. may be interpreted as a grievous insult in another country. For example, when in Moslem Middle Eastern nations, women should never wear shorts or halter tops in public. This is considered offensive in these countries and many citizens of these countries will feel both personally insulted and that you are insulting their culture.

Learning the local culture is not the only challenge of living abroad. Find out what the country's social and political problems are before you get there. Americans tend to think that people all over the world have the basic freedoms we do. As an American, you have freedom and a sense of that freedom that can only be appreciated when you live in another country whose citizens may not enjoy the basic freedoms we do. Not understanding this can lead to problems. Most Americans feel that they can do the same things abroad they can do in the U.S. Many of those unfortunates have found out otherwise when they attempted to traffic in illegal drugs abroad. Some of those Americans will spend the rest of their lives in extremely unpleasant Third World prisons.

Realize this when you're abroad: It's their country, and you will have to learn to play by their rules or you will get into a great deal of trouble. This problem cannot be over emphasized. *Obey local laws*!

Pleasantly enough, many overseas locales are safer than the U.S. In these countries the risk of physical harm is less than that in many North American cities. Street crime in many countries is minimal. But unlike the U.S., where there is virtually little terrorism, there are some countries where it is far more likely to be taken hostage than to be mugged for the contents of your wallet. Therefore, know the country where you are moving. Know the threat level in the country you will be living in.

Just what is the level of terrorist activity in the country? This is an obvious question that needs answering when moving to a new country, along with: What sort of terrorism are you likely to encounter? As you now know, not all terrorism is the same. Some forms of terrorism may not affect your company but will affect you and your family. If you are moving into a terrorist environment, get answers to these questions:

1. What are the tactics of the local terrorists? Are they kidnapping executives, assassinating diplomats or are they mainly bombing facilities? The type of tactics used is important. Knowing the methods, means and types of attack terrorists use will help you develop your personal security program. For example, if the terrorists are doing a lot of kidnapping, you have an obvious and immediate problem, one you need to address in order to stay healthy. NOW.
2. Once you have established kidnappings as a potential method of terrorist operation, you will need to analyze the threat further. Examine how they are conducting the kidnappings. Are they kidnapping people from their vehicles? From their homes? While they are at work? Determining the place and method of attack will help you develop your security program. Later in this book there will be entire chapters devoted to protecting yourself while in your vehicle, at home, and in your office. If they are attacking property and not people, are they bombing buildings, homes and/or equipment? Do they prefer arson over bombing? This could indicate that the local terrorist organization is either poor economically or not politically powerful enough or sufficiently organized to steal the explosives they need. It could also indicate a kidnapping threat to you, because kidnapping American executives is a sure method of getting funds for further operations.
3. All this gives you an idea of where the attack(s) will be coming from. Have their bombs been intended to cause death or destroy property? Did they set the bombs off when they knew the buildings would be vacant or filled with workers? Many times such off-hours bombings are used more to intimidate than kill. If they are bombing buildings on weekends and evenings, you should think twice about working late or on weekends. You could easily become an unintended victim. Are the bombings directed only towards American firms, or are European or even local companies targeted?

Your personal security program takes shape around the answers to some basic questions:

1. What are the local terrorists doing?
2. Who are they doing it to?
3. How frequently do they do it? In some countries, kidnapping is the national pastime, occurring so frequently that they go virtually unmentioned in the U.S. press. But they are probably recorded in the local press. If they are, it would be a good idea to send some local news clippings back to the home office. This will heighten their awareness of the local situation while reinforcing the need for personal protection.
4. You must determine if there is any pattern to the attacks. Do they attack after there has been publicity about the company? Or to celebrate a particular holiday? Oddly enough, many terrorist groups celebrate their birthdays by committing an act of violence. Most of us are content with a simple cake with candles. Learn if there are any patterns the terrorists follow, such as selecting one type of business over another.
5. Are they planning their attacks carefully? Examine accounts of past attacks. Have they put together a good plan? It may come as a surprise to you but if a terrorist group is well-organized, with meticulously planned attacks, this propensity for precision can actually work to your advantage. It is far easier to defend against a carefully-planned kidnapping than a random act of violence. If the act is very well planned, it will take time to develop. The terrorists will take the time to keep you under constant surveillance. This gives a good personal security program plenty of time to spot the potential kidnapping before it comes to fruition.
6. If past attacks have been successful, why have they been successful? Analyze these attacks. Find out where the victim went wrong. You will probably learn that a successful attack is a function of how flimsy the victim's personal security program was versus how cunning the terrorists were. This is because terrorists go after those with weak personal security. Past victims more than likely had a very high profile, and more than likely were kidnapped at some point in their daily routine. The victim probably kept a fairly predictable time schedule, so that when the terrorists made their final selection, the victim was easy to get. If you know of an attack that was foiled, try to find out how it was stopped, and why? What did the potential victim do to beat the terrorists?
7. What is the history of the local government in terrorist matters? Does the government have the capability to determine the threat level and do something about it? Do they have a history of helping Americans,

and what specifically have they done to eliminate the terrorist problems in their country, if anything? A country that has a good record of containing terrorism is usually a country that is very friendly with the American companies.

8. What is the stability of the country? There are some countries that never seem to be stable, and others that present the appearance of stability. A government's stability generally goes hand in hand with the popularity it enjoys with its own people. It's hard to create terrorism under a government that's popular with its people. On the other hand, a country that appears stable may only seem that way because its rulers govern with an iron hand. Such repressive governments are frequent targets of terrorism.

   The moral is: Look behind the scenes to find out what's really happening in the country. That means talking to its people.

   Another measure of the country's stability is its economy. If the country has a stable economy, it is, more than likely, a strong country.

Once you have examined all the data you can, you must make a decision about the threat level. Here is where consultants are of critical importance. They can guide you in choosing what path to take and how much personal security you will need. If you and the consultant feel that the personal security program is too restrictive, you have the alternative of leaving the country, or having your family leave the country, although this choice is seldom made. Once you have evaluated the problem and the decision is made to stay, look carefully at:

1. What can be done to reduce you and your family's vulnerability to terrorism. This requires careful study, and development of a plan of action. Seek professional help. Make sure you know your alternatives. If you have a consultant available but would like to do your own planning, you should at least have a security professional look at the plans you have developed.

2. What needs to be done to implement the plan? What will be required in the way of expenditures, manpower, and change in lifestyle to get the plan working? Be realistic. Don't avoid this problem. Face it head on.

3. If you have access to a security department, use it. Find out what help can be expected from the security department. Be ready and willing to accept and use their advice and help. Place your trust in their decisions. Don't be surprised or offended if they are not emotionally

involved. The security man is a mechanic, he knows what the problems are, and knows the cure. For you this situation is new; for the security professional, it's old hat, a job that he's done before. He has consulted with other security experts and knows what does and doesn't work. Listen to what the security pro has to say, just like you'd listen to the golf pro at your country club back in the States. Only this pro is not advising you on how to stay out of the rough, he's advising you on how to stay alive. Pay attention. Your life may depend on what he tells you.

4. How much help is needed? Does the company need to hire guards, buy security systems, acquire armored cars, etc.? Although this type of security can appear very expensive, the lack of it is much more costly. Always keep in mind that you and your family are the recipients of this security and that there's no time to quibble over costs once the shooting starts. This is not to say that you shouldn't be careful and be sure you are getting your money's worth. Get the best equipment available, equipment you or your security people know to be reliable. Then use it.
5. Will a personal security program infringe on your life style? Yes, probably. If it does so to the point that it's not used to its fullest advantage, then a lot of money and time have been wasted and you have been exposed to unnecessary danger. If you feel the program is too restrictive and may not be used by you or any members of the family, then transform the program into one you can live with. A personal security program you cannot live with creates a dangerous situation in which you feel protected but aren't. It's like carrying a gun that doesn't work. You feel protected but when the deal goes down, you're completely vulnerable. Plus, you'll have created a situation in which your opponents think you are heavily protected and they'll make sure their guns do work. An ineffective protection program often has holes in it big enough for terrorists to reach through.

THE KIDNAPPING PROCESS

The biggest threat to you while overseas is kidnapping. You must develop an anti-kidnapping program. The first step is to examine your weaknesses. Kidnappings can occur *anywhere you are not adequately protected.* Terrorists will kidnap you when and where the risk is the lowest. The risk of kidnapping is directly related to the security of the space you are in at any given moment. When terrorists plan a kidnapping they pick the point in your routine where the security is minimal. This also applies to members of your family. History indicates that the likelihood of a family member being kidnapped is small, so don't create any undue fears, but don't get too relaxed about family security. In most countries, your family is safe from direct terrorist attacks. The major threat to them is the off chance of them being hurt by being in the wrong place at the wrong time. The more unstable the local government, the more likely this can happen.

Kidnap victims may be taken from:

1. Where they live.
2. Where they work.
3. Where they visit. (country clubs, health clubs, social visits, going out to dinner, etc.)
4. While going to and from these places.

In a terrorist environment all places and activities can be unsafe—but some are much less safe than others.

Strange as it may seem kidnappers need, and often get, the cooperation of their victims. You must be easy to get, worth getting, and not very security conscious. If you are careless, with either no security or, worse yet, sloppy security, you're an inviting easy target.

AVOIDING THE KIDNAPPER

This is not as tough as it may appear. The safest thing to do is to always assume that you are a terrorist target. Here are some simple rules:

1. The first and probably the most important issue is to take personal security seriously. Don't look at it as something that is being forced upon you, and that you will do only what is necessary and nothing more. Develop a proper attitude. Realize that in certain parts of the world, personal security is part of doing business. In some circles a squad of well-armed, burly bodyguards is considered a symbol of status. You'll actually be admired for having them. You must be successful, otherwise why would you need so much protection?
2. Look for advice from people that have some experience with these matters. Don't go at it alone. Seek the advice of security consultants. They can make your life so much easier. And longer. At the very least, seek the advice of the embassy personnel.
3. Understand that you and your family will have to make some sacrifices. Personal security costs more than money. It means a change in your lifestyle. You have to be able to accept that.

4. Become a student of terrorism. Learn what the warning signs are, and have a plan ready to implement. *Look at what the other American companies are doing for their people.*

You can reduce the danger of kidnapping by adopting some simple procedures:

1. Develop a plan that allows you to avoid the potential kidnapper.
2. Put enough obstacles in the way of the terrorists so they look someplace else for their victim.
3. Be aware of your environment at all times, know how to spot changes in it. Terrorists are waiting for you to make a mistake, to drop your guard. They want to find the weakest point in your security program and attack you there. Don't give it to them.
4. The place to start with security is in the home. Make sure that your home is well protected. There will be more on this in Chapter 7, along with ways to make the workplace safer in Chapter 8.
5. You must have good communications. No matter what happens you have to be able to contact the people responsible for your protection, any time of the day or night and you must be able to contact people outside the country.
6. If you have to travel away from home, keep constantly in touch. This may require you to have good radio communications.
7. Don't draw attention to yourself.
8. Be very careful that you or your family are not enticed from a secure area into an unsafe area.
9. Change your daily routine so that it's hard for a third party to predict your movements. Leave for work at different times. If you have regularly scheduled meetings by the day, week, or month, do what you can to alter that schedule. Keep the meetings low profile. In one of the great terrorist coups of all time, Carlos the Jackal once managed to bag the oil ministers of nearly every OPEC nation during a well-publicized and poorly-guarded meeting in Vienna. The ministers were released when Carlos' demands were met. OPEC has since beefed-up security and adopted a list of security countermeasures that would make a book in themselves.
10. Resist being a creature of habit. Don't eat lunch at the same restaurant at the same time every day, jog at the same time, play tennis at the same place at the same time. Fight predictability.
11. If worst comes to worst, leave the country.

Prevention is far better than cure. Avoiding a kidnapping can become a very costly business. But the cost is far less than what might be incurred if a kidnapping takes place—especially since you may not be around to pay the final price.

In summary:

- You must be aware of your habits and routine, and you must be willing to alter them.
- You must decide what level of security you are willing to accept.
- You must develop a positive attitude towards security.

YOUR TERRORIST IMAGE

You need to ask yourself some important questions. Do you make comments that are not flattering to the host country and are you foolish enough to do this in public? If you degrade the country, you are giving terrorists ample reason to select you as a target. Even if you do have some criticisms of the country, don't express them in a way that could come back to haunt you. Respect the country and its people. Domestic workers, security guards, drivers, all must be treated with respect and dignity. That does not mean the manager has to be needlessly intimidated, but treating them with disdain invites problems. The Marighella "Mini-Manual" says that all good little terrorists need to develop information sources about their victims. A good way to do that is to have inside help. Treating the employees with disrespect makes for increased turnover and makes the terrorist's job that much easier. Treating native employees badly also provides incentive for them to hurt you by passing helpful information to terrorists.

Honestly examine and evaluate how you present yourself to the outside world. Obvious and ostentatious displays of wealth and status are inviting targets for the terrorist.

Never openly defy terrorists. Or, worse yet, challenge them. One executive we know once bragged that he carried a gun and if attacked, would not be taken easily. When the terrorists finally attacked him, they simply out-gunned him. The result was multiple and needless deaths, and the terrorists got both him and the money they were after.

Be aware of the image you present to the public. Terrorists need targets with propaganda value. It's their form of a marketing program. A company wanting to market a product researches the audience it's trying to reach, and develops a marketing program to reach them. Terrorists do the same thing; they have a target audience that they are trying to reach with their message. This audience must believe the message. Therefore what the terrorists' claim about the kidnap victim must contain some truth, or they won't be believable. So know how the local population feels about you and the corporation you represent. If the feedback you receive is negative, you and your corporation must do some local fence-mending. Good public relations, along with some philanthropic action, such as building or rennovating a city park, or similar recreation facility, can go a long way. If the general public has a positive image of you and your corporation, terrorist claims against you will be harder for them to accept.

Does all this mean that you have to spend your entire day worried about personal security for you, your family, and your corporation? No.

WHEN TO GO TO CONDITION RED

Most people today live in the category identified by security experts as Condition Green. This is typified by people so involved in their world that they are totally unaware of their surroundings.

Among other lifestyle alterations required for successful survival, you must enter into an awareness level known as Condition Yellow. Expressed simply, this means knowing what's going on around you. For instance, be aware of strange vehicles parked on your street, watch your mirrors while you drive to check if anyone's following you, know where building exits are at all times, don't sit with your back to the door in a restaurant. These precautions are not paranoia; they are a common sense approach that will enhance your perception of life by encouraging you to see and hear more of what goes on around you, while keeping you alive to enjoy this new perception.

Condition Orange goes into effect when a threat is perceived and defensive tactics are being considered. The operative word here is *perceived.* The threat does not yet have to materialize. Assume a strange noise is heard in the residence at night and you decide to investigate. A Condition Orange response would include taking a weapon along during the investigation. If the threat turns out to be unfounded, the immediate availability of a weapon has caused no harm. However, if a defensive attitude is not adopted and an actual threat does present itself, the absence of an adequate response, whether in the form of a weapon, an alarm system that will bring aid quickly, or an escape route, can spell disaster for you and your family.

Condition Red involves identifying a threat and reacting to neutralize it. This reaction can be anything from running away to engaging the threat with deadly force.

5

Case Histories of Terror Against Multinational Corporations

AMERICAN BOMBINGS

In a business school graduate study program, students learn by reviewing the case histories of past business successes and failures. You can take the same approach with your personal security program. Studying case histories of past terrorist incidents will provide you with insights on ways to better protect yourself. When you examine past terrorist incidents, you'll find that terrorists are limited to six basic tactics:*

- Bombings (Easy to do and requiring few people. Bombings account for half of all terrorist incidents.)
- Assassinations;
- Armed assaults;
- Kidnappings;
- Hijackings;
- Barricade and hostage situations.

*Brian M. Jenkins, *International Terrorism; The Other World War* (Santa Monica, CA: Rand Corporation, 1985) pp. 12.

For the purpose of personal protection you need to learn about bombings and kidnappings. While it is unlikely that you will become the target of an assassination, or an armed assault, you could become involved in a hijacking or hostage situation but that is mainly a problem of being in the wrong place at the wrong time.

First, take a look at the tactic of bombing. Bombs are either meant to kill or to intimidate. Bombs that are meant to intimidate are usually directed against facilities such as buildings, power lines, etc. These types of bombings are not usually intended to harm people, only cause damage, slow productivity and make a point. Bombing can be used for any reason the terrorists may find useful. Put another way, bombings do not have to make sense. All they have to do is scare people. As a method of producing terror, bombing works exceedingly well.

Our first case study is of a group of American terrorists, the "United Freedom Front." This group was expert at using bombing as an harrassment. It's important to note that this bombing did not take place in some faraway land, but right here in the U.S. The following is a list of bombings and the one attempted murder charged to this group.

April 22, 1976:	Bombing of Suffolk County Courthouse, Boston, Mass.
May 11, 1976:	Bombing of Central Maine Power Co., Augusta, Me.
May 1976:	Bombing of Middlesex County Courthouse, Lowell, Mass.
July 1, 1976:	Bombings of National Guard Armory, Dorchester, Mass.; empty jet at Logan International Airport, Boston, Mass.; Essex County Courthouse, Newburyport, Mass.
July 2, 1976:	Bombing of Seabrook, N.H., Post Office
October 1978:	Bombing of Mobil Oil offices, Waltham and Wakefield, Mass.
February 2, 1979:	Bombing of Mobil Oil, Eastchester, N.Y.
December 21, 1981:	Shooting death of New Jersey state trooper Philip Lamonaco. Tom Manning and Richard Williams are charged.
December 16, 1982:	Bombing of South African Airways, Elmont, N.Y.
December 16, 1982:	Bombing of IBM, Harrison, N.Y.
May 12, 1983:	Bombing of Roosevelt Army Reserve Center, N.Y.; also Navy Reserve Center, Queens, N.Y.

August 21, 1983:	Bombing of Army Reserve Station, Bronx, N.Y.
December 13, 1983:	Bombing of Navy recruiting center, East Meadow, N.Y.
December 14, 1983:	Attempted bombing of Honeywell Corp., Queens, N.Y.
January 29, 1984:	Bombing of Motorola Corp., Queens, N.Y.
March 19, 1984:	Bombing of IBM, Harrison, N.Y.
August 24, 1984:	Bombing of General Electric, Melville, N.Y.
September 26, 1984:	Bombing of Union Carbide, Mt. Pleasant, N.Y.

One of the more remarkable aspects of this group is their operational longevity. In all of these bombings, the terrorist group was careful not to hurt anyone. They went to great lengths to warn those in or near their targets. For instance, the Navy recruiting center on Long Island received a call from an anonymous source who said, "Don't talk, just listen. There are three bombs in the building for the Navy. They will go off in 30 minutes. Get out." The phone call would seem to indicate the group did not want to cause anyone any harm.

One thing to learn from the tactics used by the United Freedom Front is that you should instruct those who answer your company's phones to take all bomb threat calls seriously. The bombers are calling because while they want to wreck the facility, they don't want anyone hurt. It doesn't matter if the person answering the call thinks it's a hoax or a prank call. If it is, what's really lost? A few hours' worth of productive time, sure, but would you like to be the person responsible who refuses to clear the building? If you're going to make a mistake in a situation like this, it's vastly more preferable to err on the side of safety. Consider the consequences if you were wrong.

On January 17th, 1984 the United Freedom Front sent a pipe bomb through the mail to the Right-O-Way Air Freight Company. The package was opened by the president of the company. It exploded, but he was only slightly injured. Why they changed their tactics against this man is unknown. This same group that wanted to kill this man was previously so intent on not hurting people that when they bombed the Motorola Company they put the bomb in a package with the word "BOMB" written on it in big red letters!

As a manager for an American corporation working in a foreign land, you must realize the threat of terrorists bombing your building. Your main concern is to protect yourself and your employees from harm.

THE KIDNAPPING OF MR. HEINEKEN

Heineken planned his security program carefully, but not as carefully as his kidnappers planned their countermeasures to it. The Heineken kidnapping is an excellent case study. Heineken is the chairman of the board of the beer company that bears his name. He possesses great wealth (estimated to be upwards of $500 million), maintains a very high profile and is recognized throughout his home country. He is considered an extremely influential man in western European politics due to his close friendship with the Dutch royal family. He fits the terrorists' profile beautifully. He is wealthy, valuable to someone, and influential.

Heineken knew he was at risk prior to his kidnapping. He knew this because of what happened to his friend, Dutch industrialist, Mr. Caransa. Caransa was kidnapped in October of 1977 as he left a hotel in Amsterdam. He was released five days later after he negotiated his own release. The kidnappers were never arrested. Moreover, the ransom money was never recovered. During the investigation of the Caransa kidnapping the police found a list that included Heineken as a possible future victim. Heineken had been in the running for a kidnap but Caransa was eventually judged a more worthy target and the easiest to grab. When Heineken discovered he was on this list he took a number of precautions to protect himself, buying extensive hardware including new fences around his villa, a system of guards and dogs, and even an armored car for his personal use.

As he was leaving his office on the evening of November 9, 1983, Alfred Heineken was kidnapped as he walked towards his waiting car.

As he approached his chauffeur-driven car, he was overtaken by two masked men who began to force him towards a mini-van parked nearby. Mr. Heineken called to his 57-year-old driver for help and the man rushed to his aid. Witnesses reported that both men were beaten and forced into a waiting van. A third attacker drove away with Heineken and his driver.

The van was found a short time later, approximately one kilometer away, with the motor running. According to police records, the van had been reported stolen the previous July. The fact that the van was stolen in July and the kidnapping conducted in November indicates the kidnappers were planning the attack for quite some time.

On November 23rd the *Times* of London reported that a coded message had been placed in the classified section of a Dutch paper by the Heineken negotiators sending "warm congratulations" to the kidnappers and inviting them to make contact ". . . for practical reasons". By this time, word had leaked out that the kidnappers were demanding the equivalent of over ten million dollars to be paid in French, West German, U.S. and Dutch currencies.

To illustrate the sophistication of modern security, a magazine article claimed that Heineken was wearing a small electronic transmitter hidden in his belt at the time of his kidnapping. The device, reportedly purchased for $18,999 from an American company, was said to emit silent signals to aid in tracking kidnap victims. The article asserted that one of the kidnappers unwittingly took Heineken's belt as a souvenir and police were able to track him in that manner.

At the outset, the Heineken negotiation team decided that a ransom would be paid. In a recorded message, they were told to place the ransom in mail sacks and transport it in a white van with red crosses painted on its sides. A single driver started out in the van on November 28th on a journey that was to stretch over 150 miles before the money was delivered. The kidnappers guided the driver along the route by means of messages left in plastic cups at various locations. Finally, the driver was instructed by two-way radio to stop over a viaduct and toss the money down to a waiting truck. The money delivered, the police and the victim's families waited expectantly for Heineken's release.

It did not come.

THE LIBERATION OF MR. HEINEKEN

Increasingly concerned for the two men's safety as time wore on, the police followed leads that led them to a sawmill. Early in the morning, of the third day following delivery of the ransom money, a 70-man police team launched an assault on the sawmill. No one was found in the building and the police were about to end their search when one official noticed a lock on a rear wall of one of the corrugated-tin buildings. The lock was forced open and the searchers were met by what appeared to be a false wall. When the wall was broken down, police found a relieved Heineken and his driver chained in separate concrete cells.

The two had spent the three weeks since their kidnapping, clad only in pajamas, in unheated cells. Shaken, but without serious injury, the men changed into clothes the police had optimistically brought along for the occasion and following debriefing interviews, were happily reunited with their waiting families.

Simultaneous with the raid, police conducted a three-city sweep that resulted in the initial detention of over twenty people and the recovery of a portion of the ransom. Many of those detained were subsequently released—much of the initial confusion apparently resulted from the fact that many of the suspects were related by blood or marriage.

Initial reports indicated that the kidnapping was planned by a key mastermind who was joined by three principals, one the son of a former Heineken employee, in actually carrying out the scheme. As of this writing, two of the key players are still at large—possibly in Spain—and the bulk of the ransom has not been recovered.

Reports in the West German press speculate that Heineken may not be the only party with a stake in the unrecovered money. The December 5th issue of *Der Spiegel* reported that he was insured against kidnapping by Lloyds of London. The article that alleged the use of the hidden mini-transmitter claimed the device was a requirement of the terms of Heineken's kidnap insurance.

While it is widely known that kidnap insurance is available (Lloyds first underwrote such policies in the wake of the Lindbergh kidnapping), it is unusual to see a report on a specific case. Details of such coverage are normally kept in strict secrecy for fear of encouraging attacks against those covered. In the case of corporate clients, knowledge of the policy is usually restricted to a few key people, and the underwriters may cancel if the policy's existence is disclosed. The reason behind the surprising Heineken revelation could be the reluctance of an underwriter to renew such a policy once a claim has been made.

LESSONS LEARNED

What are the lessons that we can learn from the Heineken incident? Heineken spent a considerable amount of money on equipment, but still failed to protect himself at all times. This is a common phenomenon. Many people who surround themselves with security equipment feel that it will do all the work for them. Without the proper attitude, sophisticated equipment is just so much fancy junk. Personal security for a man who is a well-known potential target is a 24-hour-a-day job and it demands vigilance. Terrorists look for the weak link in a security system. Heineken's weak link was the area between his office and his waiting car. That's just where his attackers hit.

Most prominent people live with the awareness that they are faced by a potential for danger not shared by the average person. In the case of executives living overseas in a terrorist environment, this is especially true. Because of this danger potential, they are willing to invest in security hardware such as closed-circuit TV systems, armored cars, fences, and guards, but are rarely willing to make the kind of security requirements that require them to change their patterns, or make some similar sacrifice in their personal life.

THE KIDNAPPING OF HANNS-MARTIN SCHLEYER

The Hanns-Martin Schleyer kidnapping is probably the best example of a kidnapping of a prominent businessman and therefore, an excellent learning tool. If there is any terrorist group in the world that represented what terror against business was all about it was the German "Baader-Meinhof" terrorist group. This group was to terrorism what the Beatles were to rock n' roll; they defined the genre, providing the standards against which later groups were to be measured. Baader-Meinhof really started the kidnapping of businessmen that plagued Europe not that long ago. They decided that instead of attacking well-guarded politicians, they would select more vulnerable "capitalist figures" as their targets.

The result was that, in Europe, businessmen soon became a category of kidnap victims unto themselves. Few fit the Baader-Meinhof model better than Hanns-Martin Schleyer. He was the president of the West German Industrial Employers Federations, as well as chairman of the board of Mercedes-Benz. Like Heineken, he was aware of the terrorist threat and took some precautions. His two homes were guarded by a pair of police officers on a 24-hour basis. Unlike Heineken, Schleyer also covered his journeys to and from work. An unmarked police cruiser with bodyguards tailed him morning and night.

Shortly after 5:00 in the afternoon of September 5, 1977, Schleyer left his office in Cologne to drive home. The Baader-Meinhof group had kept Schleyer under surveillance for a long time. They knew exactly what time he would leave his office and exactly what time he would approach the kidnap area. That evening Mr. Schleyer took the usual route home in his chauffeur-driven Mercedes. Following closely behind him were three police officers in another Mercedes. At approximately 5:30 PM, a car driving the wrong way on a one-way street cut in front of Schleyer's car, forcing his car off the road. The Schleyer car hit the curb. As the attackers' car cut in front of Schleyer's vehicle, a baby carriage rolled into the middle of the street. This was a ruse, intended to block the street in the event Schleyer's driver decided to try and drive around the car that blocked the road. The terrorists knew no one, no matter how desperate, would run over a baby carriage. Anyway, the issue was made moot when the second car, containing the bodyguards, and was following too closely, crashed into the rear of Schleyer's car.

Five terrorists who had been waiting in a parked Volkswagen van nearby walked up to the bodyguards' car and fired their weapons point-blank into the bodyguards. The three police officers never had a chance. They made a heroic effort to return fire but were simply outgunned. A passenger in the vehicle that cut off Schleyer's car walked up to the driver of that car and killed him at point-blank range, being very careful not to hit Schleyer.

Schleyer was taken from the Mercedes and put into the Volkswagen van. An indication of the planning and organization involved in this ambush and kidnapping was the fact that it only took about 90 seconds to force two cars to a stop, kill four people, kidnap the target, and begin the getaway.

At approximately 8:00 p.m. the van used to take Schleyer away was found. Inside the van was found the first message from the kidnappers, addressed to the German federal government and threatening Schleyer's execution.

In this case, the terrorist group did not want money, they wanted concessions from the West German government. They wanted the release of 11 imprisoned terrorists within days, 100,999 German Marks for each of them; use of a plane to fly them from the Frankfurt airport to a country of their choice; televised broadcasts by the released gangmembers; and the placement of two known left-wing liberals on the flight to insure safe passage.

The gang's message was acknowledged by a brief message broadcast on the same evening during the 8:00 news, but the kidnappers' demands were not met. The government used a radio broadcast to ask the terrorists to give them a sign that Schleyer was still alive. The terrorists used a clergyman as a go-between with the government, and provided a video tape of Schleyer, along with a demand it be broadcast.

The authorities refused to televise the tape, and the terrorists responded by a telephone call to the press in Dusseldorf, threatening "consequences" for Schleyer if this demand was not met. With the kidnapping of one man, the terrorists were in a position to dictate much to the national and international press. The police refused to broadcast the tape.

The end result was tragic. Mr. Schleyer was killed on October 8th. This abduction was and is still considered a classic. The terrorists did not want money, they wanted the release of their comrades. They meticulously organized the kidnapping. They watched Schleyer's movements and planned the exact location of the ambush. They probed the security team's weak points, uncovering the bodyguard's lack of firepower. They followed Marighella's outline perfectly. They used the tactic of surprise perfectly. The driver of Schleyer's vehicle had no idea of what was happening. They took total command of the situation by being brutal and quickly killing everyone in the entourage except Schleyer.

When examining these two cases, one of the first things we can see is that different terrorist groups can operate in vastly different ways. The American group made efforts not to hurt people but to destroy property, except for one attack. The Baader-Meinhof gang felt no such compunction against killing, which they did coolly, methodically, and as a central part of their strategy. The killing was not done in response to an attack from an unexpected direction; planning called for the permanent elimination of Schleyer's bodyguards and eliminated they were.

The disparity between terrorist group methods is why knowing as much as you can about the group operating in your area is so important. The type of group sets the level and intensity of security you will need, as well as dictating the appropriate response to them. You can separate the groups into:

Amateur kidnappers: Appearing all over the world at an alarming rate. These are mostly criminals trying their hand at kidnapping. They are very dangerous, not because of their professional abilities, but because they have a tendency to panic and kill their victims (or let them die) when negotiations run aground, or even when they do manage to get their hands on the ransom money. These criminals are fond of leaving their kidnap victims wherever they have been holding them when the ransom money becomes available. In many cases, they've been known to abandon their hostages in places where, if the police had not accidently found them, they probably would have perished from thirst, cold, hunger or sheer neglect. Amateurs tend to consider a live captive as a dangerous liability. Moreover, these groups are plagued with problems in assessing their own risks correctly.

Professional criminals: By comparison, display a much more reasonable risk-assessment capability and tend to let their victim live, *if their demands are met.* In case of apprehension, they hope this tactic will win them a more lenient sentence. Police stalling tactics tend to have a much less negative effect on these kidnappers than on amateurs, who may suddenly see their dreams of quick riches shatter and therefore abort a kidnap operation, killing the victim to cover their tracks. On the other hand, if the original demands are not met, professionals might resort to such drastic blackmail methods as those encountered in the Empain case. Baron Empain, a forty-one-year old Belgian multimillionaire, was kidnapped outside his residence in Paris on January 23, 1978. He had left standing instructions that in the event he was kidnapped, no ransom was to be paid. Initially, his family respected this, but on the third day the kidnappers cut off the little finger of the Baron's left hand and sent it to his wife. The Baron was held for sixty-three days in three different hideouts, the last one a tent in the cellar of a suburban house in Paris, where he was chained hand and foot and usually kept hooded.

The Baron's family co-operated with the police throughout the kidnapping and helped maintain a news blackout following the initial reports of the kidnapping. The initial demand called for a ransom of $20 million. After two months of negotiation, the family agreed to pay $8 million. By then, the police were ready to act. The man who delivered the ransom was a police inspector. On March 23, the inspector followed a treasure hunt of clues but the kidnappers made no contact, presumably using the dry run as a way of satisfying themselves there was no police surveillance. Fortunately, the police were discreet enough not to let themselves be spotted and the next day, when a real ransom dropoff was attempted, they ambushed the site, killing one kidnapper and capturing another. Three men escaped.

The captured kidnapper was thirty-nine-year-old Alain Caillol, one of the gang's leaders. Son of a successful manufacturer, he lived in a luxurious villa with a swimming pool, managing a branch of his father's firm, while leading the double life of a criminal.

The penalty for kidnapping and murder in France is death; and Caillol realized if the Baron was killed, he could be guillotined. He therefore telephoned the rest of his gang from the police inspector's office, warning them that their only hope of escaping the guillotine themselves was to release the Baron at once. On March 26, Easter Sunday, the Baron was dumped in the outskirts of Paris, and took the Metro to the centre of the city before telephoning his thunderstruck wife to come pick him up. The police located the gang's hideout, and further arrests were made.

This case illustrates one of the more successful procedures in dealing with a criminal kidnapping. The victim's family, who could well afford to pay a large ransom, were patient and co-operated with the police. The police allowed them broad latitude in negotiation, and showed admirable discretion and skill in ambushing kidnappers at the ransom drop. Competent handling of the case after the ambush led to the recovery of the victim alive and in good condition, without payment of a ransom. The existence of the death penalty strengthened the hand of the police, working almost with the force of conscience to keep the kidnappers from killing.

Overall, professional terrorist groups have shown a tendency "to be true to their word," that is, to release their victims when their demands are met or kill them when faced by a flat denial. The Baader-Meinhof bunch proved this in the Schleyer case, and it is certain that if the West German government had bowed to the terrorists' demands the gang would have released Schleyer. This reliability is due mainly to an inherent characteristic of all terrorist actions: a desire to be recognized as an opponent in the struggle for power, a "partner in the game." Their risk and security assessments are usually within reasonable bounds, and they consider their victim more a tool in their war against the authorities than as a human being. A note of caution, however: if they don't get their way, these groups are also ruthless and will not hesitate to do away with their hostage.

In the Schleyer incident, it is evident that terrorists will not shrink from a major gunbattle when they are after a target they consider worthwhile. In general, you can protect yourself with hardware but large scale safety precautions, such as those taken in the "hardening" of political target figures, are extremely costly and beyond the resources of any except the wealthiest men and biggest corporations. Today, good protection costs a great deal of money. An armored car costs anywhere from $85,000 to $225,000 dollars depending on the equipment and the level of protection needed.

ASSASSINATION OF RAFAEL TRUJILLO

The May 30th 1961 rubout of Dominican Republic President Rafael Trujillo is a fine example of what not to do in a politically violent Third World environment. Although Trujillo was a political leader and not a businessman, a look at his assassination presents us with a good example of a thorny problem: that of protecting a prominent public figure who leads a secret double life. The plan behind the Trujillo assassination was a simple one. As with nearly all terrorist actions, it was based upon his daily routine.

It was an open secret in the Domnican Republic's capital city of Santo Domingo that Trujillo regularly visited his mistress who lived on the outskirts of that city, taking the same route to and from her home every time.

Along with the advantage of a victim religiously keeping a regular schedule along a regular route, the assassins had inside help. One of the conspirators was located in Government Palace and informed his colleagues in plenty of time for them to get ready for the assassination, well before Trujillo's departure from the palace.

When the message finally went out that Trujillo was leaving the palace, three vehicles prepared for the attack. One would be located off of the highway waiting for Trujillo's Chevrolet, a vehicle well-known to Santo Domingo residents, to pass. The first car would then swing out in pursuit of the big Chevy, blink its headlights to alert two more cars waiting up ahead, and at the same time open fire on the Trujillo vehicle.

The attack began as soon as Trujillo's car appeared. The group fired their weapons into his car as it passed. Bullets shattered the rear window and Trujillo was wounded. Riddled with bullets, the big car screeched to a halt.

Trujillo leapt out, firing his revolver. In a moonlit firefight, Trujillo was hit, fell to the pavement and died.

This case study highlights a very delicate issue. Examine your activities carefully. If there is something you do that you would like to keep secret, that's fine. It's your business. But keep it a secret from everyone, especially terrorists. A word of caution. You may find this far more difficult than it sounds.

Trujillo had an activity that he kept secret from *almost* everyone, except the terrorists. During the second phase of target selection, terrorists watch your every move. Be aware that while you may be keeping a secret from some people, you may not be keeping it secret from everyone. Never, never, never hide anything from your security department. If you are leading a "second life," tell your security department. Don't worry about being embarrassed. It's far better than being dead.

THE KIDNAPPING OF ALDO MORO

Aldo Moro is another classic case, and one we have mentioned before in this book. Although somewhat similar to the Schleyer kidnapping, the Moro case is different in some very significant ways.

In March 1978, the kidnapping of Aldo Moro, respected elder statesman of Italian politics, shocked the world. The impact on the Italian consciousness was comparable to what the reaction in the U.S. would be if ex-

Presidents Ford, Nixon, or Carter had been kidnapped. All of Italy came to a standstill.

On March 16, 1978 shortly after 9:00 AM, after having stopped at a nearby church for communion, Moro was en route to the Italian Parliament in his dark blue Fiat 130. He went to the same church, at the same time every morning, establishing an easy-to-follow pattern. Each and every morning you could count on Aldo Moro being in the same place at the same time. Also with him in the Fiat were a chauffeur and a bodyguard. Three security guards followed in a white Alfa Romeo, in a configuration similar to that of the Schleyer abduction.

As the cars moved through Rome's fashionable hotel district and approached an intersection, a car bearing diplomatic license plates pulled ahead of the Fiat and stopped suddenly in the intersection. Moro's chauffeur applied his brakes so abruptly that the backup car hit Moro's Fiat.

The driver and passenger from the blocking car got out as if to check for damage. Approaching Moro's car from both sides, they pulled out pistols and shot the driver and security guard in the front seat of Moro's car, killing them instantly.

Prior to the shootings, four men dressed in Alitalia airline uniforms had been standing at the intersection as though waiting for a bus. As the action commenced, they crossed over to the cars, pulling automatic weapons out of their flight bags.

The four gunmen fired at the security officers in the follow-up car, killing two of them immediately. The third security officer rolled out of the car onto the street and was able to get three shots off at his attackers before he was neutralized by a fatal shot from a sniper on a nearby roof.

A blonde woman and her male companion stood at the intersection and appeared to be in charge of the operation.

Moro was transferred, along with a briefcase containing official documents and another containing medications, to another blue Fiat.

The obvious planning and attention to detail in this kidnapping was remarkable even when compared to the precision tactics of other kidnappings.

For example:

1. The escape route involved entering a restricted street which was closed off with a padlocked chain. The kidnappers were prepared for this, carrying bolt-cutters.
2. A simulated auto accident nearby diverted traffic from the attack site.
3. Police were diverted from the attack area by false bomb reports in other areas.
4. The telephone system was apparently out of order immediately following the attack in the district where the kidnap occurred.

5. The stationing of a sniper on a rooftop resulted in the death of the only officer who had a chance to react. This was a stroke of genius. A gunman in such a position could command the entire field of action, providing cover and/or neutralizing any unforeseen opposition. Many kidnap and robbery plots foul up because they are too rigid; no one is stationed in such a way as to be free to be "creative," to have the freedom of action and the means to handle unpredictable developments in an improvisational way. The use of this rooftop sniper is exactly the sort of "wild-card" unexpected threat that makes security people go grey-haired before their time.
6. The diplomatic license tags on the decoy car had been stolen from the Venezuelan Embassy more than a year before the attack.
7. A street flower peddler usually set up his business at the corner where the kidnap occurred. That morning, he found the tires of his station wagon slashed where it was parked near his house. This kept him home, and out of the way that day at the scene of the attack.

In this case, as in so many others, Moro was not the initial target. The man the Red Brigade originally wanted to get was travelling in an armored car, his bodyguards following him in another armored car. This man changed his routes all the time. His bodyguards were well armed. During the final target selection process, the terrorists decided that this target was too hard to get and would result in a dangerous, costly attack. They moved on to Moro, the weaker target.

Following the Moro kidnap, the searches, apprehensions and trials that followed rocked Italian society to its roots. A major offensive was waged against the Red Brigade and the group was assumed to have long been wiped out when it surfaced again with the assassination of an American. Red Brigade members ambushed Lemon Hunt as he drove up to his home. Mr. Hunt was the director of the multinational force in Eqypt's Sinai Peninsula. As he sat is his car waiting for electric garage gates to open, terrorists opened fire. The car was armored but the terrorists knew that. They also knew where there was a weak spot on the armor and that's where they concentrated their fire. Hunt was killed. Again, homework through surveillance paid off and the terrorists found their target's weak point.

For you, the lesson is clear. Know and identify your weaknesses and make them strong.

THE KIDNAPPING OF MR. NIEHOUS

On the night of Friday, February 27, 1976, seven masked men carrying guns burst into the Caracas, Venezuela home of American businessman William

Niehous, tied up his wife, an American maid, and the family's three teenage sons, and kidnapped William Niehous. Niehous, 45, was vice-president of Owens-Illinois of Venezuela.

On the following Monday a local newspaper printed a manifesto from the kidnappers stating that Niehous had been "arrested and imprisoned" pending a people's trial for interfering in the internal political and economic affairs of Venezuela. The group said they were not interested in ransom and that Niehous would be immediately executed if the police took any repressive action against leftist groups.

Police located an abandoned car used in the kidnapping as well as some clothing Niehous was wearing when abducted. Mrs. Niehous claimed that three young men she had hired to cut grass around the house a few days earlier were among the kidnappers. Two of the three were identified as escaped guerrilla revolutionaries.

The kidnappers identified themselves as the "Comando Argimiro Gabaldon," believed to be a splinter of the Venezulean "Bandera Roja" group. The kidnappers sent statements to the authorities that "the fulfillment of three demands will largely determine the course of the trial of Niehous which we are carrying out" and reiterated the threat to summarily execute Niehous if there was police repression against the populace or if the police discovered the group's hideout. The guerrilla group issued a list of demands.

1. That their political manifesto be published in newspapers in Venezuela, New York, London, and Paris.
2. That Owens-Illinois distribute 1,200 food baskets in Venezuela.
3. That the company pay each of its 1,600 Venezuelan employees a $116 bonus. Total: $185,600.

The threat to Niehous' life had to be taken seriously, partly because the guerrillas sent photographs to newspapers showing Niehous standing beside a hooded, armed guerrilla. The group also threatened four politicians with kidnapping if Owens-Illinois and the Venezuelan government did not give in to their demands.

On Tuesday, April 6, after more than six weeks of fruitless searching for Niehous, a political manifesto from the guerrillas appeared in newspapers in New York, Paris, and London detailing the "crimes" of Owens-Illinois against the people of Venezuela. The statement revealed that the company had paid its workers a 500 bolivar bonus. The guerrillas expressed their solidarity with other revolutionary groups such as the ERP in Argentina and MIR in Chile.

Shortly afterwards, the Venezuelan government took steps to nationalize Owens-Illinois' subsidiary for breaking national law, undermining the Venezuelan constitution, and bringing the government into disrepute by negotiating with the guerrillas, giving in to their demands, and publishing a statement produced by a subversive group.

The Niehous kidnapping well illustrates the complexity of a single terrorist action, how a single action can carry far-reaching repercussions up to the very highest levels of local governments, as well as on the victim's corporation.

The Niehous' abduction followed the usual tactical plan of a terrorist kidnapping. Precisely why Niehous was kidnapped is somewhat less clear; beyond the reasons stated by the terrorists, however, lies a thoroughly expounded revolutionary strategy for the kidnapping of American business executives. In the Niehous case, as in the other, often tragic cases listed in this chapter, there is much to learn from.

THE NEED FOR RESIDENTIAL SECURITY

There are four basic environments you need to protect: your office; routine travel, such as to and from your office; leisure activity; and your home. From your family's point of view, the most important place to protect is your home. Although kidnaps of people right from their home is unusual, it has happened. William Niehous, Vice President of Owen-Illinois of Venezuela, was taken out of his home in Caracas, Venezuela and held in captivity for 40 months. U.S. Army General James Dozier was also seized in his home in Italy and held hostage. While your home may be your castle, as far as the terrorists are concerned it's far less fortified than a castle, and sometimes an ideal place from which to kidnap you.

Personal security begins at home, protecting you and your family. If you are living in a foreign country, or are about to move to a foreign country, an important part of your personal security program is selection of a secure house or apartment (houses, however, tend to be far more defensible) in a secure location for your place of residence.

LOCATION OF RESIDENCE

It's an ancient adage, but still relevant. The three rules of real estate are location, location, and location. If you are about to move overseas, and you are in search of a home, location is the first and foremost priority. The home should be located in a safe section of town but not the section of town that houses only the very wealthy if at all possible. Living in the most exclusive section of a Third World city is a luxury that you will generally find affordable. But living among the very rich will only attract undue attention to yourself. Unfortunately, in many countries you may not have an option on the type of area that you live in. In some countries there are only homes for the very rich or the very poor, with nothing in between. In such situations there is no option; you will have to live in the wealthy area. This is a fringe benefit for Americans working overseas, but bear in mind that terrorists are always looking for signs of wealth.

The layout of the local road system around the home is important. The home's location should be an area that has easy incoming and outgoing access. The ideal home permits alternate routes to work. The worst possible situation is a home in an area with just one road in and one road out. As we have learned, terrorists need to fix a time and a place to conduct the ambush. If there is limited access to the home, it is very easy for them to pick the time and a location to conduct a kidnapping. Statistics simply show that most kidnappings happen while a person is in a car on his way to work. There is an obvious explanation for this. If there is limited access to your home and you have only one road to drive out of every morning, and you leave your home at roughly the same time every morning, it becomes child's play for terrorists to pull off a kidnapping.

DOGS AS PROTECTORS

No matter where a person lives, a dog is one of the best security devices money can buy. Admittedly, dogs are hard to obtain in many countries, but they're worth it. Dogs are amazing alarm systems, and almost certainly the first alarm system known to man. Despite the fearsome reputation of breeds such as German Shepherds, and Dobermans in the attack role, any terrorist that's reasonably quick with a firearm can neutralize that threat with the flick of a finger. Dogs are best as warning devices, not as weapons. They have a much better sense of smell and sound than many animals their size. This keen sense of smell and hearing allows a dog to warn you that someone is trying to get into the home. This can buy the occupants of the home some valuable time. It may be just 30 seconds, but in an emergency situation those 30 seconds can make all the difference in the world.

If the dog is barking at the door, it's telling the people inside that there is something it considers suspicious outside, allowing the owner to investigate and determine what the problem may be. Almost any dog will do, but the bigger the better. Although barking may be annoying at times, the louder and more often the dog barks the better. Big, aggressive dogs can be nice pets, and the most loyal friend you're ever likely to have, but along with ownership goes responsibility. If an untrained dog gets loose it can cause a great deal of harm.

There are dogs that should be kept in the house and ones that should stay outside and protect the house perimeter. The dog kept outside is used primarily for keeping people away from the yard, and as a warning dog to a lesser degree. The problem with this is that you may not be able to hear the dog barking out in the yard. If your objective is to keep terrorists away, a barking dog in a yard is not much of a help. His warnings may go unheard and he probably won't last too long in a terrorist situation anyway. We are dealing with people who think nothing of snuffing out the life of a human being, killing a dog will not faze them in the least.

The best option is a house dog, it can give you warning of intruders coming into the house and will slow them down once they're in the house. If the terrorists are concentrating on a noisy, aggressive dog, it may give occupants more time to develop a defense and/or escape.

It's not a nice thing to think about but you must plan a course of action if someone tries to get into the house. While your dog is bravely buying you time, you should already know what to do with the time the dog is giving you. Plan ahead for what you will do while the dog is slowing the attackers down. Map out various escape routes. Which you take may depend on the method of attack used. The amount of time you have will depend on the type of dog that you have.

DOG TYPES

A watchdog is a fourlegged burglar alarm. It can warn you but that's all it does.

Attack dogs fall into one of two general categories: either a deterrent that attacks on your command, or one that attacks anything that is not familiar to him. If the dog's owners are not around and the mailman is not familiar to him, then that dog will attack the mailman. If the decision is made to have an attack dog in the home, then you and the company are assuming a special responsibility. If the dog decides to chew on the mailman or eat one of the neighborhood kids, you will be responsible. An all-out attack dog will buy you more time, but will certainly require more caution and attention on your part. Do not take attack dog ownership lightly. It can be rather like keeping a loaded gun around the house.

Never needlessly scare people with your dog. The dog becomes excited and confused for no good reason.

HOUSE GUARDS

Having human guards around an American home 24 hours a day is virtually unheard of. Home security has a different meaning abroad than it does in the U.S. Stationing guards in your home is not a pleasant thought, but in many countries it is a very common practice. And a necessity.

There are two preferred methods of hiring reliable guards. Some companies employ their own. Company-employed guards must be trained in home security. Many companies prefer to hire former or off-duty police officers. In many countries, police officers are permitted to hold two jobs.

Just because a guard is a present or former police officer doesn't mean he's qualified to guard a home. Police officers receive little or no training in protecting private homes. Their training is how to react to public situations. In private situations this training can prove detrimental. Police officers are trained to help people in need. It could be easy to distract a police officer by setting up a ruse, such as a bogus accident or crime victim. To be effective, your guards must understand that their job is to protect you and the house, not to help other people. For a good cop, this can be a tough lesson to unlearn.

Rather than hiring your own guards and training them, many companies will hire guard companies to supply the service for you. If you are part of the company's decision-making process when it comes to hiring outside guard services, be sure of their capabilities.

In the home, these guards are the first line of protection. Therefore they should be treated as professionals and with respect at all times. No one in the family should mistreat or needlessly distract them, or use them for anything else except what they are paid to do: guarding the house. If they carry weapons, they should obviously know how to use them. Don't assume that just because a guard is carrying a gun he knows what to do with it, and keep in mind that along with knowing *how* to use their weapons they must also know when. A jittery, trigger-happy guard can be a big problem. Such an improperly-trained guard is a liability not an asset. If possible, use two guards, one inside the house and the other outside. Good security guards are the centerpieces of your security team. You or your security department must constantly keep your guards updated as to what you're doing and what you expect of them. The mere sight of a guard that looks as if he knows what he is doing may be reason enough for terrorists to eliminate you from their potential victim list.

DOMESTICS

Maids and servants are another affordable luxury when working overseas. However, like guards, in many countries domestics are no luxury. In many ways a maid can provide better security than a guard. A guard is a good deterrent in the event of an attack, but domestics can add a whole other dimension to home security. Due to their familiarity with the country and neighborhood, loyal domestics can become good sources of intelligence data.

Planning is the key to any organized terrorist action. Planning requires surveillance. Surveillance requires terrorists to watch the home. Terrorists hanging around the neighborhood watching your home means more unfamiliar faces in the neighborhood. An observant maid can probably recognize an unfamiliar person, or a suspicious situation that other occupants of your house would never be aware of.

This is especially true if you are new to the country. A sight that may look normal and completely unremarkable to a newcomer in the country may shout for attention to someone who has lived in that country all their life. Domestics' knowledge of the country and its way of life can be invaluable to your family. Your economic status in a Third World country may allow you to hire more domestics than just a maid or two. You may have gardeners, drivers, cooks, houseboys, etc. Make them all part of your security scene, making sure to know their strong and weak points. They can be your eyes and ears. Use their experience to enhance your security program. Keep in mind that they must be treated with respect at all times, and treated as professionals. Your family must earn their loyalty. Loyal domestic help to the family can be your family's best protection.

Since your domestic help will be spending a great deal of time in your home and with your children, you must subject them to a thorough and extensive interview and background check. Go over each person very carefully. These people are going to be with your family more than you will. You want honest, reliable people in this position. Use your local U.S. Embassy as an intelligence aide in this. Embassy staff may have records of locals who can and cannot be trusted. Local police can also be helpful in this role, but not always. Staff at the U.S. Embassy will be able to tell you, with a fair degree of accuracy, whose side the local police is on.

Marighella's terrorist training manual suggests that terrorists try to infiltrate the victim's environment through domestics. Be extremely careful if you're hiring.

In the U.S., it is hard to check on the background of employees but in many foreign lands it is as easy as checking their criminal record with (trustworthy) local police. Ask for and check out *all* references. The family member who will be spending the most time with that domestic should interview the person and be completely satisfied that they are comfortable with them. It is wise to have both the husband and wife agree to their employment after separate interviews. Do not force an unwanted domestic on your spouse. As superstitious as it may sound, rely on your "sixth sense." If something or someone seems odd or unsafe, go with that impression and select someone else. The ideal situation is to hire someone who has worked for a friend or an acquaintance of yours, someone you know to be reputable. Keep in mind that you are allowing someone into your home who knows when you will be out, has access to all the keys of the home, who will know as much about you as anyone in your family. They must be trusted so they must be worthy of trust.

TELEPHONE SAFETY

The family and everyone associated with the family needs to know secure methods of using the phone. Proper phone use and the monitoring of unusual phone calls is very important. There is no reason to get overly concerned about a few random odd phone calls. Every strange call is not a terrorist looking to attack you at home, but the phone is one of the easiest ways to get a fix on where the family is located. Phones have historically been used by terrorists to gain information about you and your family. One basic precaution is the unlisted number, a basic security precaution that all executives should take advantage of no matter the country they are working in. Anyone with teenagers knows that controlling the home phone is difficult if not impossible. But this is the real world, full of real dangers, and not American suburbia. Impress your family that these security precautions are real, that the threat they're meant to confront is real, and that you are not putting them through such inconvenience just to entertain yourself. You must exercise control over your home phone number, even to the point of keeping it out of the company phone directory.

No one in the family should identify themselves to a phone caller. If the phone caller does not know who you are, why is he calling you? Don't identify the residence, don't identify the phone number. Wait for the caller to ask for a family member by name, and never tell a stranger on the phone who you are. If the caller appears to have the wrong number don't give him yours. Ask him what number he is calling, and ask him to try it again. Never give your name to a stranger on the phone. Just because someone dialed your number does not mean they have your number written down. They could have dialed the number randomly and are hoping that you give them the number. The scenario can go like this:

The caller: "Is Bob there?"
The answer: "There is no Bob here."
The caller: "Who am I speaking to?"
The answer: "This is Mary."
The caller: "What number is this?"
The answer: "You have reached 555-1234."

The phone call should go like this:

The caller:	"Is Bob there?"
The answer:	"There is no Bob living at this number."
The caller:	"Who am I talking to?"
The answer:	"What number are you looking for?"
The caller:	"Ah, 555-1243."
The answer:	"Sorry you have the wrong number, why don't you try again?"

A home that's constantly getting wrong numbers, or other unusual calls needs to have an organized effort by everyone in the household to log those calls. Family members will have to record the time of the call, what the caller wanted, along with the sex and apparent age of the caller. Nuisance calls, such as obscene phone calls, should be reported to the phone company, and your security department at once. Do not argue with callers or even try to talk to them, just hang up. If callers are threatening any household members, take the calls seriously, do not take them as pranks. Immediately call the local police, your company security people and the nearest U.S. Embassy or consulate.

Don't be tricked into leaving the house. Know who you're talking to. If you get a phone call from the local police, or a government agency make sure they are who they claim to be. In any country, you, the company and company personnel have an obligation to cooperate with the police. That does not mean that any member of the family must unquestioningly follow the instructions of someone who represents themselves as police officers, especially over the phone. Make sure you positively identify the caller as a member of the law enforcement agency he claims to represent. This can be done by telling the caller that you cannot stay on the phone right now but you will call him back within a few minutes if he gives you the phone number. Then check the phone number against the telephone directory to make sure it's the right number. If you can't verify the number, call your security department right away. One of the most frightening calls you can get is that one of your children or your spouse has been hurt in an accident, and that you are immediately needed at the hospital. In a situation like this, you'll likely act first and think later. It is critical that you take a moment to consider the situation. Ask the caller for the name of the hospital, and check with that hospital before you leave home. Take a moment to check with the police as well, just to verify if there has actually been an accident.

You might take the extra step of assigning family members a secret code word, known only to you and them, to verify emergency communication. While not a foolproof method, (no security technique is 100% foolproof) a code word could save time in an emergency.

Keep in mind that in many parts of the world the phone systems are stone age throwbacks, and that wrong numbers may not be uncommon. Nevertheless, take the time to check out who you're talking to and whether what they're telling sounds kosher.

Keep all important numbers near the phone. Obviously, the standard emergency numbers such as police, fire department, and hospital should be there. But everyone in the family should know how to contact the company and its security department any time of the day or night. This includes any of the domestic help. Make sure they know how to call the security department, and who they are talking to when they do call.

SECURITY WITH DOORS

Your first consideration with the construction of the home is the doors. It doesn't take much to get through most doors, just a couple of good, solid kicks is all it takes.

In the "good old days" doors were made of solid wood but, while new construction techniques have reduced the cost of doors, they have also reduced their effectiveness against forcible entry. Many of the doors built today have wood or glass panels. These panels of glass or wood can be knocked out quickly, and more-or-less silently, by whoever wants access to the home quickly and quietly. Attackers can get into your house by merely breaking the glass or wood panel, reaching in, and unlocking the door. The best thing to do to a door with these sort of panels is replace it. Ask your company's security department, or security consultants for advice. But if neither is available, and you are forced to make these decisions by yourself, the best replacement is a hollow metal door. These doors can be very ugly, but they are very strong. Some of the heavier models can stop low-power slugs. If you need less-than-maximum security, and, looks matter, a 1¾-inch solid-core hardwood door will do almost as well.

Strengthening doors applies to every door in the house. It's not wise to beef up one door when there are two, three, or more outside doors that are also vulnerable. A terrorist, or just a good burglar, will have a fair idea of where the strong and weak points are in each house under consideration for an attack. Such an intruder is not likely to accept the professional challenge of getting in through the security door over an assured entry that's quick and quiet by way of another door.

Sliding doors require special attention; they are the most difficult to secure. Most of these types of doors are made of glass, with useless locks. To make them more secure, install shatter-proof glass such as Lexan® or other impact-resistant material instead of glass. Also install a vertical lock in the floor and ceiling, or get a broom handle and put it in the track that the door slides on. Even if the lock was forced open it would still be impossible to get the door open.

The bottom line is simple exterior doors should be designed to keep both unwanted people and unwanted things out of the home. Any home anywhere in the world that houses the most prized possessions the executives have, their families, needs to be designed to keep unwanted people out.

LOCKS

A good door is only as good as its lock. There are hundreds of different types of locks. The locks that are presently on your doors were more than likely installed by the builder. These locks are known as primary locks. Locks installed after construction to increase security are called auxiliary locks. The primary lock is usually a key-in-the-knob or spring-latch type lock. These are common, good-looking, and inexpensive. But not very useful. It's common knowledge that a credit card can be used to slip the latch or any good wrench used to twist the handle off, after which it's a simple matter of disengaging the lock and getting into the house.

Chain locks are popular but not very effective. They are great for telling burglars that someone is in the house, and that may be enough to convince him to go someplace else. But that's no deterrent for a kidnapping attempt. If you have chain locks and are attached to them, they should be well-anchored with long screws or bolts. In no case should chain locks be used as a primary lock on any home.

There is another type lock called a "deadbolt". It offers more security but still can be opened with a crowbar. Every house should have a deadbolt lock on the doors, and preferably two deadbolt locks. The bolt on the deadbolt lock should be at least 3/4" long; a 1" bolt is even better. The safety of the home hangs by a fraction of an inch when someone makes a determined effort to force his way in.

Where possible, have a "roller bar" inserted in the bolt. This is a small bar of metal that can rotate freely within the dead bolt. If someone is able to expose the bolt, and attempts to cut or saw through it, the bar spins when the cutting tool contacts it. Because it spins, the blade can't easily bite into the bar, which makes the work of sawing the bolt off an extremely time-consuming and frustrating experience. Terrorists and burglar alike do not have time to spare. They want to hit and run, so anything you can do to slow them works to your advantage.

Locks should not be so good at keeping terrorists out that they keep you from entering your house. No matter how many locks are on the doors, or what type of locks they are, they should not take so long to open they can endanger people legitimately trying to get into the house. Imagine standing outside your door, fumbling with keys, unable to either get inside while being pursued or to retreat if someone comes your way. Speed and ease of entry for yourself and all other authorized personnel are essential to a good home lock system.

Another type of lock that is frequently used is the double cyclinder lock. This lock requires use of a key any time the door is opened—inside or outside. That makes it difficult for a thief to get out of the house with your goods (or a kidnapper with you) if he doesn't have the key. But it also means that should you surprise an intruder inside your house and are trying to escape, you'll need both the time and presence of mind to unlock the door. Double cyclinder locks also have disadvantages in a fire. You may be overcome by smoke before you can find the key and open the door.

All locks should have protective metal plates mounted over the cylinder section. A good bonded locksmith can do it. But double-check him. Make certain he uses one-way screws or lag bolts at least 3 inches long in installing this or any other lock device.

DOOR HINGES

Something as simple as the hinges on a door is often overlooked. Hinges need to be examined.

Surprisingly, hinges are often mounted backwards, with the hinge pins on the outside of the door. You can lock and bolt the door with the finest hardware made, and if the hinges are on the outside, a kid from down the block can use a screwdriver to loosen the pins on the hinges and lift the door right off the frame.

Any door with the hinges exposed should be remounted. Immediately. If this isn't possible, the door should be taken off and a small screw hole drilled in the hinge pin and barrel of the hinge at a spot where it can't be reached when the door is closed. A pin or metal screw should be put in the hole so that the pin is anchored and cannot be pulled out.

Your door is your first line of defense against both the common and uncommon criminal. It's important that you never allow anyone to enter your home or apartment just behind you. A good way of preventing this is to make sure that any entrance door used regularly should be fitted with either a spring or pneumatic quick-close device.

Peepholes are another item that must be on every outside door. And they must be used, not only when looking to see who rang the bell, but every time you leave to see if there is anyone waiting silently for you. Remote TV cameras are getting so inexpensive that it may be a good idea to put in a video camera or two, allowing you to see what is going on outside of the house.

If necessary, two peepholes should be installed at different levels so every member of the family can easily use them. Peepholes are unusable if someone has to stoop to waist level or stand on a stool. They should be wide-angle peepholes, which allow a large field of vision when looking through the hole.

The rule about peepholes is a good rule to apply to any piece of security equipment. If a device is inconvenient, no one will use it.

Since no door or lock is any more secure than the keys which open it, keep keys apart from anything that identifies who you are. Do not have your name or address on the key chain. Anyone who laid hands on your keys would also be thoughtfully provided with your name and address, and be thoughtfully presented with the opportunity to go into your home and loot it.

Because it's sometimes necessary to leave the car ignition key with a parking attendant, that key should be kept on a separate key ring from keys to your home and office. Parking attendants have no legitimate need for the trunk key; don't leave it. Never give anyone your house keys for any reason. Remember: modern keymaking equipment can duplicate keys in a matter of seconds.

WINDOWS

If most modern doors are vulnerable to terrorists, modern windows might as well not even be there. When the door is locked, the window is the next-best way to get in.

Getting in through the window is as American as apple pie, motherhood, and the flag. At the same time, it's as un-American as a terrorist or psychopath trying to get into the house. At the very minimum, you should install a large nail or bolt in the window track in such a way that it prevents the window from opening wide enough to let someone in. If you use the window for ventilation you should install window locks. Windows that you don't use for ventilation you should lock, or even nail shut so they can never be opened.

Windows will always be vulnerable for the simple reason that no matter what type of lock you put on the window, it's impossible to prevent someone from just breaking the glass and unlocking the lock. Therefore, install shatter-proof glass in all your windows.

LIGHTS

Intruders need time to get into your home. Darkness gives the time they need. In most conditions, a minimum of four lights on each corner of the house are needed to properly light the outside. Any areas that remain in shadow can be used as hiding places. Make sure all areas are well lit. Lighting inside the home is obviously important. When you walk in and flip on the nearest wall switch, you should be able to light up the entire room. You should not have to blunder about the room in darkness switching on separate lights. Put some lights, and even a radio, on timers to give people the impression that someone is home. The radio is especially effective in this role. Few prowlers will enter a house in which music is playing. The radio may be a ruse, as far as they're concerned, or it may not. But the criminal that's willing to experiment to find out won't be in business long.

Keeping drapes closed is important. Sheer curtains are trouble for a security program. Drapes that you can see through are also trouble. Hang thick, opaque drapes or blinds, through which you cannot be seen from outside of the house.

FURNISHINGS

Furnishings aren't really a security issue, but need to be examined from a safety standpoint. All your drapes, rugs, and carpeting must be flame retardant. Check your furnishings for polyurethene. The nicest thing to say about polyurethane furniture is "Get rid of it." When this type of furniture burns, it emits deadly hydrogen cyanide in the form of extremely dangerous noxious fumes.

Each room in the house should be equipped with a fire extinguisher. By this I mean *all* rooms. There may be rooms in which a fire extinguisher will clash with the decor. Be ingenious and find ways of hiding the fire extinguisher. No matter where it is, be sure everyone in the family, especially the domestic help, know the location of each extinguisher, and how to use it. Make sure the extinguisher is one that will put out electrical, oil, grease, wood and paper fires.

Don't forget smoke alarms. They should also be in all rooms if possible but at the very least in hallways, near bedrooms, the kitchen, and the garage.

Don't try to put out major fires by yourself. Make sure that the local fire department number is kept beside each phone.

INTERNAL SAFE HAVENS

The thought of a safe room is another unpleasant subject, but if you are living in a terrorist environment, and serious about survival, it's definitely something that needs to be considered. A safe room is a room in your house which does not allow access to it from the outside or from any other place in the house. It can be a bedroom with additional locks on the door or a specially built room within the home used for this purpose only. The purpose of the safe room is to protect you and your family from an attack by people who entered your home by force.

In most homes the sleeping quarters are separated from the rest of the house's living spaces. In a terrorist environment, consult with your security people and install a reinforced door in the hallway separating the sleeping quarters from the rest of the house. If all the bedrooms are upstairs, the door can be mounted at the top of the stairs. The purpose of the door is to deter an intruder from entering the area where you and your family are sleeping, and allow you time to call the police.

This inner door can be used as an indicator to tell you something about the intruder(s) and his(their) intentions. If they attempt to break through the inner door there should be no doubt that they are attackers and mean you physical harm. In such a situation, you are safe to assume that drastic actions on your part are appropriate.

ALARMS AND WARNING DEVICES

Alarms don't stop people from entering your home. Alarms let the intruder know that their presence has been detected. Alarms are good at stopping burglars but not very effective with people who are in your home to kidnap or people intent on causing you physical harm. Before you buy any burglar alarm or warning device consult your security department. They are experts and will install an alarm that suits your needs.

If terrorists realize you have an alarm system, they may consider moving on to their secondary target. If they do decide to enter your home, the alarm system will announce their presence not only to you but to neighbors and hopefully the police. This will make their planning somewhat more difficult and may persuade them to choose an easier target.

If you have a security department, rely on their advice on alarms, but remember the best alarm system available presents no real physical barrier. Alarms are simply warning systems, a psychological barrier, and a method of calling for help. An alarm can never replace locks and barriers.

If you have to plan the alarm system by yourself, make sure all doors and windows on the ground floor are alarmed, along with all windows and doors on upper floors that are accessible from the roof, balconies, gutter drains, and trees. Each floor should have an interior trap installed, such as pressure sensitive mats underneath the carpeting, passive infrared detectors, or any other device that will warn when someone is in the house. If you hire an alarm company, do it through your security department. It must be a reputable company, since it's employees will have all the information necessary to break into your home undetected.

COUNTERSURVEILLANCE

All through this book, we've been drumming the concept of surveillance detection as the single best way to protect yourself, your family, and home from terrorist action. No attempt to kidnap you will be made without some sort of surveillance of you, your home and your activities in and near the home. Although most terrorist kidnappings take place in a car, the surveillance leading up to that kidnapping starts with your home. Terrorists will be noting your departure times, and your initial driving routes from home to work. Thus, it is vital that you can detect surveillance.

Surveillance of your home can take many forms. We've run through some of them already, but here are some hints that you, your family, and domestics should look for:

Be aware of strangers, whether they are walking, driving around the neighborhood repetitiously, or simply sitting in a parked car, van, or truck.

Be suspicious of anyone requesting information or trying to start conversations with you, family members, or domestics. Pay close attention to workmen or peddlers near your home. Check out telephone repairmen, utility workers, and other workmen by calling their local office. When making notes of suspicious people or occurrences, gather as much data as possible: descriptions, license numbers, date and time, how often they are seen and patterns.

Monitor nearby vacant houses or apartments which may be used as surveillance stations.

Avoid putting any personal or descriptive material into your garbage. To the right person, it will read like a book.

If you feel that you may be under surveillance, contact the police or other appropriate authorities immediately. In most cases, if a threat group feels its surveillance has been detected and the authorities informed, it will immediately change or cancel its plans.

ACCESS TO YOUR HOME

Don't let anyone in your house. Many times a stranger can use explanations that sound legitimate to get into your house. Instruct everyone not to open the door until they have verified the need and identity of the individual. Whether it is a policeman, repairman, utility man, real estate agent or other plausible sounding caller, ask first for identification, which should be slipped under the door, then call their local office for verification, BEFORE YOU LET THEM IN.

If a stranger wants to use your phone due to, say, a car breakdown, make the call for him. DO NOT LET THE STRANGER IN. Be suspicious and use your head. Again, make sure everyone in your household is instructed in this area and follows your specific guidelines.

ABSENCE FROM YOUR HOME

In an overseas situation, absence from your home is best handled by leaving someone behind. This may be a friend, domestic, or guard. In any case, your home security is best protected by having an individual stay in your home while you are away. If this is not possible, then there are several precautions you can take to PROTECT the security of your home while you are away. Do not start by making certain that all deliveries are cut off (newspapers, mail, etc.). This is a common practice by people going on vacation. But, if you tell the people that deliver to your door that you will not be around for a couple of weeks, you have just told a random group of complete strangers that your house will be unguarded for two weeks. It's a far better idea to leave a key with a neighbor and have that person receive all your deliveries and secure them until you return. If you are a two-car family, make sure that you don't leave the second car parked in the driveway for two weeks. Use automatic timers or photoelectric switches to turn lights and a radio off and on. (Twenty-four-hour timers with variable settings and multiple hookups are ideal.) Ask your neighbor to check the premises, pull the drapes open and closed, park his car in your drive, or keep your yard trimmed. Never leave a light on twenty-four hours a day or keep your drapes permanently closed. This does little more than advertise that you are away. Only your neighbors and a trusted friend should know this. If applicable, ask the local police to check your house occasionally*.

When you return, use caution before entering your home. Check for signs of forced entry or unusual activity. If things do not look quite right, go to a telephone away from your home and call the police. Then you can both check out the house.

We've just gone through quite a few domestic security precautions and you're probably telling yourself, "Good grief, if I have to do all that to my house, it'll end up looking like a maximum security prison. Which of these devices and modifications do I really need to do, and which can I safely do without".

The answer to this question depends largely upon you. Unless you have an expert to give you personal advice, you will have to decide just how far to carry the recommendations I have made. If you have done a good threat analysis, you should have a good idea of what you are up against, and just how secure your home should be for adequate protection. Make it as secure as you feel necessary. Make your judgments based on the facts.

APARTMENT SECURITY

Many times you may not be living in a residential home. Apartments are becoming more common for executives working overseas. The disadvantage is that apartments have limited access in and out of the building. They do offer other advantages. With an apartment, you can be in a building that has a 24-hour security guard, who announces all guests and checks all visitors.

*Patrick Collins, *Living In Troubled Lands* (London, England: Faber and Faber Limited. 1978) pp 70

The security-conscious apartment building should include some other features as well. The garage should have a controlled access. A key should be needed to get into the garage. The garage must be well lit and patrolled by security guards, if possible. Elevators are particularly dangerous areas in an apartment. Don't ride in elevators with a stranger. If someone you are suspicious of gets on the elevator, get off at the next floor. If you think something is about the happen to you, press the alarm button. Hard.

Whether in an apartment or your house, use common sense and practice good security. It is inexpensive insurance against random violence, and could possibly make a terrorist group bent on kidnapping consider another target.

7 Protecting Your Overseas Plant or Office

INTRODUCTION TO OFFICE SECURITY

Attacks on offices rarely occur, but as you beef up your home security you must in turn increase the level of security at the office. Terrorists will attack the weakest point of your security, and they can attack at any time and any place. They have demonstrated a unique ability to find the Achilles heel of a protection plan. The basic goal for a plant and office protection program is to make it as difficult as possible for the terrorists to get into the building or into protected areas

Office security begins at the point of entry of the building. The objective is to control who enters the facility. The problem is that security cannot be so tight that it restricts operational needs of the company. While it's imperative to have good office security, and keep out the people that cause harm, it is also imperative to continue functioning as a business. If security is so tight that you cannot continue to do business, it's like holding yourself hostage.

The office is far easier to protect than the home, or the trip to and from work. At the office you are surrounded by people. Most of the larger overseas companies have a security officer present at all times. If you feel that the threat level is high, there should be a security officer stationed outside your office, controlling access to that office. In the U.S., a security officer sitting outside an executive's office is not a common site. But in many South American countries it's as common as seeing a secretary seated at a desk in front of an executive's office. You must be accessible to the people who work with you but you must be careful who you let into your office. Keep in mind that you are a symbol. The reason a group may attack your office is because an attack there creates the appearance that they are attacking capitalism. Their ideology holds that capitalism causes oppression so when they attack your office they are fighting oppression. Not all terrorist groups have adopted this tactic. The best way to find out if you are susceptible to this type of attack is to examine past history. If terrorist groups in your country have attacked American offices and office personnel in the past, it's a good bet they'll do it again. It's a good idea to get serious about office security.

Work with your security department to organize a good security program for your office. Your security people may make suggestions that cause you some daily personal inconvenience, but you have to set an example for the rest of the company and be receptive to everything that they suggest. As always, a final decision must be made on a balance between good security and restrictive security.

The problem is that you or any member of your family can become a target simply by your presence in the overseas office of any American company, the office does not necessarily have to be yours. If your office security is good and has forced terrorists to look elsewhere for victims, fantastic.

Congratulations. After doing all that hard work, just imagine how foolish and frustrated you'll feel if you become an involuntary target in another American company's office, a company that hasn't done it's security homework as well as yours has. This is especially true if you're doing business in a country that has had bombings directed at the American diplomatic community or against American business. In countries with this problem be extremely cautious about entering buildings that are not well protected.

OFFICE LOCATION

As with the location of your home, the office needs to be located in a safe area. Unlike a residence, it's to your advantage to have an office located in the heart of the business community, especially in an exclusive or high rent area. These areas have superior security and are usually carefully watched by the police.

If your company is renting space, take offices on the upper floors of the building. If possible try not to occupy the entire building. If you are the sole occupant of a building, destroying the building hurts only your company and no others. The same reasoning holds true in locating your offices on a busy main street. The more people in and around your building, the less likely terrorists will try to attack it. At the same time, try to stay away from traditional trouble spots, such as universities and government buildings. These are frequent locations for demonstrations, and sometimes, riots. Don't let your building or the building your office is in become a convenient terrorist target.

BUILDING SECURITY

Office security begins at the outer perimeter; the first line that must be crossed by an intruder. Natural and structural barriers like chain link fences, gates, and shrubbery will be the first obstacles faced by attackers. These barriers need to be as terrorist-proof as possible, while remaining functional. Barriers need openings to allow access to the facility, but every opening in the perimeter barrier is a potential security hazard. Obviously, these openings must be held to a minimum, keeping in mind that security must be balanced with other safety considerations such as fire regulations. Barriers cannot be so good that the workers cannot get out of the building. Any possible way into the building needs to be guarded, whether it's windows, doors, roofs or common walls. All must be protected against entry by use of guards and/or electronic surveillance equipment. If you are located in an area where car or suicide bombing is a favored method, you should consider limited vehicle access to your building, specifically in the area of underground garages. A well-placed bomb in an underground garage or under a modern building in which the first floor consists largely of supports holding the rest of the building up, can bring the entire building down. This was how a light truck packed with explosives massacred over 240 U.S. Marines in the infamous 1983 Beirut barracks bombing. Under these conditions your security personnel must make it as difficult as possible for a vehicle to get within close range of the building.

At night, all buildings need protective lighting illuminating all of the exterior surfaces of the building. The lighting needs to be designed in a way to sufficiently deter intruders and make intruder detection certain. It's proven fact that all types of criminal including terrorists avoid well-lit structures. To have lights, you need power. Even during the day it's impossible to conduct "business as usual" without lights. In most modern buildings, it's easy to destroy a building's power source and plunge it and everyone in it into darkness. To combat this problem, it's vital to have an auxiliary power generator to make sure the lights stay on in the event of a power failure. In some countries disrupting the power is part of the daily routine, as a sort of noon whistle for the entire city. In areas with a terrorist/guerrilla problem, the local power generating station is a popular bombing target. Electricity is a sometime thing in the Third World.

Inside the office a protective system of guards and/or electronic devices should be implemented. This needs to be done without interfering with the orderly and efficient operation of the business. Of course the easiest way to keep out terrorists is to allow no one to enter the building, and, of course, this is impractical. Access control must be maintained. All employees must wear identification badges. Guards monitoring all the entrances must check the badges of all the employees anytime they enter the building. Due to the fact that almost all large companies, whether overseas or in the U.S., routinely employ this type of employee identification and screening, employees are usually not the source of terrorist action. Visitors become the biggest problem. All visitors to the office should be required to identify themselves and state their business, a common procedure in the U.S.A., and one that should be religiously followed overseas. The identification badges for the visitors and employees should be tamper-resistent and bear a good, clear photo, perhaps a photo taken in front of a unique backdrop that would be difficult for I.D. forgers to duplicate. Any identification system is only as good as the enforcement employed within that system. As a manager overseas you should set an example and insist that everyone in the building wear their badge at all times. Obviously that includes you. No one should be allowed to roam around the office without authorization. Security should challenge anyone in the building not wearing an I.D. badge, even if they are known employees and escort them back to their place of work. Visitors without company I.D., should be politely booted off the property. Visitors should receive "visitor badges" and be escorted wherever they go.

As with domestics, anyone who works in the building and are not company employees must have security checks conducted on them. You must know who you have wandering around your building. This especially includes the maintenance people who clean up at night, who, for the most part, work without any direct observation. Maintenance staff are the only people to have access to your office without anyone watching. It is imperative you know who they are.

Keep in mind that, like maintenance, the members of the cleaning crew have their own keys to the offices, work unwatched at night, and have access to all parts of the office. Make sure the security department has properly screened these people. If the building is not owned by your company, get assurance from the owners that these people have been checked out thoroughly. In high threat areas, the only realistic solution is to hire your own janitors and check them out yourself. It is sometimes best to have them do their work during the day when the office staff can be around to watch them. This may be an inconvenience, but well worth the security benefits.

Office and plant security is a science unto itself and the nuts-and-bolts portions of the operation should be left in the hands of security professionals who understand the details of providing good office security. However, in many cases you may have to make recommendations or take an active role in developing systems yourself. In that event, some key issues need to be addressed.

- Everyone entering the building must be screened in some way and register with a security guard or reception.
- All visitors must be provided with badges and escorted to wherever they are going. They should never be left alone to wander through the building. This includes even your personal friends. It's vital for you to set the example.
- Executive personnel files and records must be kept confidential. No one except those with a need to know should ever have access to these files. The information contained in them could prove quite useful to a terrorist group.
- Those with a need to know and access to executive personnel and family records must be instructed that they are to be kept strictly confidential. This includes home addresses or telephone numbers.

OFFICE TELEPHONE SECURITY

Good phone security starts with the secretaries, but not only the secretaries should be trained in good phone security. Anyone who uses company phones must be instructed to never give out any information about you or any other employee. They should never tell any caller what time any employee might be expected to arrive to or return from work.

Remember at all times that terrorists need to know all the pertinent times and places of their potential victim. Nobody in the company should give information that will tell the caller where and when anyone in the company will be. Not giving out company information over the phone is a simple matter of training. Well-trained secretaries will not have to be told, it is something that they do automatically. But just to be on the safe side make sure that the secretaries know who and what they can and can't talk about over the phone. There is one phone-related problem for which almost no amount of training can prepare personnel: the bomb threat. It is probably the most terrifying experience a secretary can have. Brief secretaries and other employees responsible for handling the phone on ways to handle bomb threats and other violent threats against the offices or its employees. Role-playing games, in which one employee is the phone answerer and the other the caller can be useful. As in any emergency situation, the purpose of good training is to instill in the trainee a plan of action well-known before the actual emergency, so that when the emergency becomes real, that person will react swiftly and calmly.

EMERGENCY PLANNING FOR BOMB THREATS

Bombings represent the greatest threat to American offices overseas. So it follows that offices abroad need to have some form of bomb detection equipment. This may sound like a drastic measure, but you might be surprised how many companies in the U.S. already have bomb detection equipment in their offices. This equipment should be selected and purchased by a security professional. Do not try to do this on you own. For complete protection against bombings, you must also have an emergency evacuation plan. The plan can be as simple as any standard fire evacuation plan. As the person responsible for the safety of your workers, you should look over any existing plans and make sure they do not create panic. If your company is located in an area that has a problem with terrorist bombings, the merest mention of a bomb in the building can trigger panic. Workers running around in a state of panic can be worse than the bomb itself. Panic, defined as a "sudden, excessive, unreasoning, infectious terror" is one of the most contagious of all human emotions. In the context of a bomb threat, panic is the ultimate achievement of the caller. Once a state of panic has been achieved, the potential for personal injury and property damage is increased dramatically.

One way to control panic is to have a plan. You must make sure that there are people in your facility who have been trained in evacuation procedures and know what to do. Everyone in the company must know who can authorize an evacuation. Designate someone who is nearly always on-site as the person authorized to call an evacuation. You must select this person promptly and it must be somebody who can make the right decisions, someone you have absolute trust in. Bear in mind that terrorists will not cooperate and make sure they plant the bomb in the building when this person is there. No one can be on-site all the time, so you must develop a chain of command made up of managers you trust to know when to call an evacuation. The conditions under which they can authorize an evacuation should be clear-cut. Have the security department develop the evacuation plan.

Once your employees are out of the building, your problems are not over. Somewhere outside of the building should be a command center with phone lines installed, and a complete set of emergency supplies. When creating an evacuation plan there are some basic considerations. Consider the possibility that the terrorists actually *want* you to evacuate.

An even greater danger may be waiting outside for the employees. When organizing an evacuation plan, make sure that employees be evacuated to an area known to be both safe and secure. As soon as the threat message is received notify the police of the bomb threat, and at the same time tell them that you are evacuating the building. When the police arrive they should be told where the employees are located, and asked to provide protection for them until the incident is over.

HANDLING THE BOMB THREAT

History shows that bombing is by far the most common type of terrorism. Bombings are meant to accomplish certain goals. Bombs can be simply symbolic in nature, or they can be meant to destroy the building or they can be meant to destroy the people in the building. Bombs also have another use. They can be hidden within the building and can either be set to go off at a specific time, or be set to blow up on command. If the terrorists have the ability to hide the bomb and control the detonation they can use such a bomb in an extortion attempt.

The caller has certain knowledge or believes that an explosive or incendiary has been or will soon be placed and is calling in an attempt to minimize personal injury or property damage. The caller may be the person who placed the device or someone else who has become aware of such information.

The caller may want to create an atmosphere of anxiety and panic which will, in turn, result in a disruption of the normal activities at the installation where the device is purportedly located.

What happens if you or your secretary receives a phone call and the caller says "There is a bomb in your building. It will go off in an hour. Evacuate the building immediately"? The first thing you both should do is remain calm. This requires some training. All staff members must be instructed on what to do in the event of a bomb threat. There should be a practice drill to insure that everyone knows what to do.

Remaining calm is easier said than done. Most people are not used to being told that they are about to be blown up. This is the main reason for the education program.

The person receiving the call must try to keep the phone caller on the line for as long as possible. Bear in mind that there is little that can be said to the bomber that will make things worse. The bomb is already set to go off. There is nothing you can do to change that reality. In fact, the phone call can be used as a tool in defusing the situation, if not the bomb itself. The more information the person receiving the call can get, the better. The person answering the phone can say things like "There's no bomb in the building. Who would be stupid enough to put a bomb in this building?" in an attempt to get the caller to reveal something about the bomb they may not say without being provoked.

Whoever receives the call must ask the right questions. Ask where the bomb is. Try to get the caller to reveal where the bomb is located. At this point, there's a good way of determining if the bomb threat is a hoax. Ask the caller "Is the bomb in the . . . ?", and pick a fictitious location, a place that does not exist. If the caller says, "Yes, the bomb is there," it's a good indication that the bomb threat is a hoax. At the very least this technique gives security teams a place to begin their search.

The person receiving the call should never take it upon themselves to decide there is no bomb in the building. That decision must be left up to professional technicians whose job it is to find the bomb. Of course the most important fact to ascertain is when will the bomb go off.

Other techniques for handling bomb threats:

- To keep the caller on the line as long as possible, ask him to repeat the message. Record every word spoken by the person.
- If the caller does not indicate the location of the bomb or the time of detonation, ask for this information.
- Inform the caller that the building is occupied and the detonation of a bomb could result in death or serious injury to many innocent people.
- Pay particular attention to peculiar background noises such as, motors running, background music, and any other noise which may give a clue as to the location of the caller.
- Immediately after the caller hangs up, the receiver of the call should report to the person designated by management to receive such information. Since law enforcement personnel will want to talk first-hand with the person who received the call, that person should remain available until they appear.
- Why was the bomb planted and who planted it? This may not be useful information at the time but it can help in the future.
- Who planted the bomb? This may seem like a stupid question but if the bomber is talkative he may slip up and give his name. Asked this question, the caller will certainly divulge the name of the organization responsible. Indeed, that will probably be one of the reasons for the call.

AFTER THE CALLER HANGS UP

The next important step is to notify the authorities. This is another element of the bomb threat policy you must have in place at your company. With this policy, one of your people should be assigned to be the bomb threat officer. This person makes the call and ideally is someone from the security department.

Impress upon everyone involved that this is serious business and they should all take any threat seriously.

MAIL BOMB

There is a particularly devastating type of bomb that can easily be introduced into the office. It appears small and innocuous, yet it can kill and maim, at the same time creating widespread terror. This is the mail bomb, or letter bomb. The mail bomb is used the world over as a device of terror. Mail bombs are hard to detect and can be made small enough to fit inside a business size envelope. Although they are small, they can carry a deadly payload, even killing someone who is standing a good distance away from explosion. Mail bombs have become increasingly efficient since the introduction of a powerful plastic explosive called C-4. As a plastic explosive, C-4 can be cut, molded and shaped into nearly any form the user desires. Sort of an off-white color and resembling common window-seal putty, C-4 is an explosive of unprecedented power.

C-4 is usually manufactured in sheets, generally measuring a foot wide and quarter-inch thick. Coupled with a small electronic detonator connected to the envelope flap, a business-letter-sized sheet of C-4 would tear apart the person unfortunate enough to open such an envelope.

Thanks to its pliability, C-4 can also be molded into any shape and placed inside everyday objects. Enough C-4 could be placed inside a small object such as a radio, to bring down a good part of an average-size house.

The very saddest thing about mail bombs is that they never seem to kill the person they are directed at. Most companies have mailrooms, and it's there that these bombs usually go off, killing mailroom personnel. Business executives seldom open their own mail. They have their secretary open their mail for them, and therefore it's the secretary that's killed. A mail bomb sent to the home can be even more devastating. An unsuspecting member of the family or household staff could open the letter or package in a room containing other people. Therefore it is your responsibility to see to it that not only the employees of the company but your family and household staff receives warning concerning mail bombs. What's truly frightening is that there is no real way of guaranteeing that mail bombs will not be detonated in the office. The only thing you can try to do is reduce the risk of such explosions. Again, you must rely on the help of a security professional. They are well versed on the equipment available for mail bomb detection.

You should not rely on equipment alone. There are some actions that can be taken to help reduce the risk. Train those who handle the company mail on what to look for, and what they should do once they feel they have found a potential mail bomb. The other safety measure that must be taken is for you, working with your security department, to develop a procedure examining all incoming mail. This could be a monumental task in some companies, and something that may not be possible. There are some things that can be done, however. All incoming mail must go to a central location. This is done in most cases anyway, but care should be taken that no mail is directly sent anywhere in the company without first going through the mailroom. Someone in the mailroom should be made accountable for security procedures. All too often, mailroom staff feel checking all the mail in a search for letter bombs is just another nuisance, and that security people are just paranoid. To counter this, management has to fully endorse the procedure set up by the security department. If there is no security department, and the procedures were set up by a consultant the same level of cooperation should come from management.

OFFICE ACCESSIBILITY

As always the terrorist will take the path of the least resistance. The offices most likely to be targets of terrorism will be the ones directly accessible to the public, the ones they can get to easily. Your office should be hard to get to and, if it at all possible, not located on ground floor levels. If the local threat level is high, the office windows that face public areas should be curtained and reinforced with bullet resistant materials. Direct and immediate access to your office should be monitored by a secretary, guard or other individual who screens *all* persons and objects entering your offices. Again this may appear to be harsh but in many countries bullet-resistant windows are as common as window shades. The door to your office should be a solid core door with dead bolts. If at all possible, you should have more than one way out of the office. If you have the privilege of a private bathroom it should also act as a safe room. The door to the bathroom should also be a solid door with a dead bolt. It could also be a solid metal door, for added safety.

At *no time* should an executive consider the use of a weapon in his office. First of all, the terrorists are almost certainly there to kidnap you, not kill you. The weapons they carry are primarily for purposes of intimidation, and defense. True, terrorists routinely use weapons to kill bodyguards, chauffeurs, or anyone else who tries to stop them, but as the target, they will not shoot you. Unless you force them to, by attacking them with a gun.

Face it. What are you, most likely armed with a six-shot pistol, going to do against three or four extremely agitated terrorists carrying rapid-fire automatic weapons? Die in a blaze of bullets, that's what you're going to do, and maybe injure or contribute to the death of innocent bystanders in your own office. There's just no percentage in fighting back at this stage.

ALARM PROTECTION

Your office should be equipped with a hidden and unobtrusive means of activating an emergency silent alarm. This alarm can sound at the security or guard center, and another executive's desk, or any other location where a signal would summon immediate aid.

The important part of this "panic button" alarm is that if and when it goes off, everyone knows what to do. Security personnel must get to your office quickly and quietly.

You must think before you push the button. If there's a man sitting across from you in your office with a .357 Magnum pointed at your chest, and this charming gentleman has just informed you that you are being kidnapped and that you and he will now quietly leave the building and that he will have that impressive piece of hand-held artillery hidden under a raincoat slung over his arm and trained on the small of your back during the entire walk out of the building . . . then you should probably think twice before punching the button that's going to bring security into your office and perhaps force our well-armed friend to do something that will make everyone sad.

In other words, use your head before sounding alarms. In some cases, such as the one we just described, you may not want help, at least not right at the moment.

VISITOR CONTROLS

To prevent the above scenario from happening, you must have the approaches to your office well controlled. Employees can help by questioning any unescorted people they don't recognize. There should be an electronically-controlled door between your office's main reception area and the general office area. The door should be controlled by a guard sitting in the office area behind bullet-resistant glass. Consider the use of a metal detector in high risk areas. A detector is invaluable in screening packages and other objects being carried into the executive area.

Guards should also ask all visitors to leave their coats in the reception area. Visitors should not be allowed into the office area unless guards can see what they are carrying. As we've seen, an overcoat over the arm can hide a variety of weapons.

Security guards will be faced with a variety of visitors. The expected visitor is the easiest to deal with. Guards should ask him who he's there to see, who he represents, and what time he was to arrive. His information should match the information the guard should have been previously provided with.

Unexpected visitors should be asked their business. If they appear legitimate, the person they want to see should come out to the reception area. There's some debate in security circles on whether the person sought by the visitor should come out to the reception area. If the visitor is armed and has kidnapping on his mind, his victim has just been delivered to him. The best solution is to bar all executives from seeing visitors without an appointment.

Delivery people are a special problem. Regular deliveries made on a day-to-day basis present little threat. It is the unexpected delivery that presents a major problem. Guards should not allow the deliverer to leave a package unless he knows who it is from, who it's for, and who sent it. Guards should be extremely suspicious of any delivery man who wants to drop off the package and quickly leave.

AFTER HOURS ACCESS

Admittance into your office during nonworking hours must be carefully controlled. As mentioned before cleaning and maintenance personnel must be escorted by a guard or supervisor, as well as any other persons requiring nonworking hours access into the office. In addition, guards should be directed to make periodic checks of executive office areas in their after-hours inspection tours.

All restrooms on the floors where executive offices are located (as well as all others in a multistory office building) should be locked to eliminate unrestricted public access. This is especially true if the executive offices are situated on a floor accessible to the public. Rest rooms are a favorite hiding place for terrorist bombs, or terrorists themselves, hiding out to pull off a little after-hours mischief and mayhem. Any doors leading to spaces that can hide things or people such as janitorial and other maintenance closets should be kept locked at all times. This includes doors to telephone and electrical equipment rooms. Such rooms should be kept locked and access given only to maintenance and telephone personnel who have a requirement for such access. Access to keys for these doors needs to be tightly controlled with only the supervisory and screened and approved technical personnel allowed access to these areas.

The best security situation possible is for high-risk executives to enter from a private entrance that is monitored, either by closed-circuit TV or a live guard. You may not relish this kind of scrutiny but in some places in the world, it's mandatory. The time that you arrive to work is a known quantity. It very seldom varies (although it should, for your sake). The place that you exit the car to enter the building is probably always the same. That puts you at a location at a known time. This is what's commonly referred to as a bad move. As experience has shown, when terrorists know the times and locations of a potential victim, the situation becomes very dangerous and the potential for kidnap high.

In the office, your first line of casual defense will more than likely be your secretary. It's not your secretary's job to keep someone from forcibly entering your office, that's a job for security.

But if something like a threatening call comes in, it will be the secretary who will handle it. Your secretary should remain calm and try to get as much information from the caller as possible. Secretaries should never give out any information such as executives' home addresses, phone numbers, names of family members, where they go to school, how old they are, or any other personal information. The secretary should never discuss executives' travel plans or any other travel information. Above all, secretaries should never discuss security precautions that have been taken or security equipment that's been installed in the office. That could defeat the purpose of the entire exercise. Just as terrorists depend on the element of surprise to gain the upper hand on their victims, so is security equipment and countermeasures needed to surprise their intended targets.

If a phone caller is especially persistent, secretaries should be polite but never give any information. Instead they should say that the executive will return their call.

Secretaries should also be trained in how to recognize and handle mail bombs, since they, more than likely, will be the ones opening the executives' mail.

VEHICLE SECURITY: INTRODUCTION

Every morning you willingly enter the most dangerous part of your day. The drive to and from work is by far your most vulnerable time for kidnapping. Terrorists prefer this part of your daily routine because it is so easy for them to select the location and conditions under which they'll make their move. Despite many of your precautions, it is during this time you are most vulnerable.

You may have state-of-the-art technology in alarms and protective systems at home, and the same may hold true for your office, but when you are traveling to and from work, the number of people and amount of security equipment you can access and use easily is limited. Vehicle security is one of the most vital parts of your personal security program. There are two areas of vehicle and driving security that need to be examined.

The first and most important is prevention. Prevention is the way you can avoid becoming selected as the final target of kidnapping. The second area of vehicle security is the reaction portion. Reaction is recognizing the fact that you're about to be attacked and knowing how to drive out of the attack.

Prevention is the more important of the two areas and requires a little planning on your part.

If you have done a good job of protecting yourself at home and at the office, terrorists will then move to find answers to the question of how secure you are in your vehicle. In many cases, terrorists already have a pretty good picture of your vehicle security. They've discovered this by keeping you under surveillance for several days or weeks, in part during your daily ride to and from work. If they feel your vehicle security is good, and they will have a hard time kidnapping you from your car, they will more than likely seek someone else as the final target. If the terrorists look at your home security and see that it works, then they'll turn to your office security. If it is working, they'll then go over your vehicle security. If it works, the terrorists run out of options. Why should they waste their time with you and risk failure when there are other far easier targets available?

By having good personal security, you haven't stopped them from kidnapping someone. You will never prevent a terrorist kidnapping; you will only prevent them from kidnapping you, diverting the kidnappers to other targets on their list.

The basic intent of your prevention program is to stay away from areas where you and your vehicle could be easily struck at. Don't forget the obvious. Try to stay away from all areas of the city that are dangerous. In many cities this is impossible, because it will severely limit your mobility. In many Third World cities the dangerous, poverty-stricken areas of the city are so vast there's no way to get where you're going without traveling through them. Use caution when driving through areas you know to be rough.

There are some basic safety rules you should always use whenever and wherever you are driving, no matter where you are. Simple techniques such as wearing a seat belt and keeping your windows up and doors locked at all times must become automatic reactions every time you get into a vehicle. If your threat level is high, you should consider swapping your vehicle with that of a friend or associate on a random basis. Some companies have their executives switch cars regularly with other people in the company.

SURVEILLANCE DETECTION

A big part of your vehicle security program is the ability to detect the surveillance. Detecting surveillance is the best method you have to avoid becoming the final target. If terrorists feel that you are too difficult to grab, it may mean they will divert their kidnapping to someone else. Always take surveillance seriously. If you detect surveillance, report it immediately to the police and to the U.S. embassy.

You do not have to be either a James Bond or live in a world of paranoia to detect what you think is surveillance. Not all surveillance teams are super spies; in fact, there are very few really good surveillance teams. Most terrorist organizations rely on you to be oblivious to what's going on around you and are confident you will not spot anyone following you or watching you.

Detecting surveillance is not difficult. It is a matter of knowing what to look for, being alert at all times, and not believing in coincidence. For example, if a beat-up old Chevy with two young-looking kids sitting in it suddenly appears parked in front of your house and you see the same vehicle with the same people behind you in traffic on the way home from work on another day, that may be a coincidence. But if you are having supper at a restaurant one evening soon after that and you see the same vehicle parked outside, it's almost certainly surveillance.

REACTING TO A VEHICLE AMBUSH

If you have a good prevention program you will not need to use the reaction part of a vehicle security program. However, as the Boy Scouts say, be prepared.

The first challenge in a vehicle attack is realizing that what you see in front of you is an ambush. The simplest thing to do is consider it a threat.

When driving your car, take some basic precautions:

- *Become familiar with all possible routes to and from your place of work.* Know them well. And drive them unpredictably. You do not want to become a creature of habit, so predictable that terrorists can look at their watches and confidently predict "He'll be here in two minutes." Put yourself in the terrorist's shoes. Ask yourself, "Is this an area on my route to work that I cannot possibly avoid?" If it is, then it is a likely spot for a kidnapping. Know the shortest routes to the police stations, hospitals, fire stations, or the protection of your company's plant or office building from any given point on your route. You need to know these short cuts because in the event of a problem you'll want to get to safety as quickly as possible. If you are attacked while in your vehicle, plan on staying in command of the situation. Terrorists rely on taking complete command of the kidnapping by way of a ferocious assault intended to kill or neutralize the target's defenders and shock the target into a submissive mindset—any action that slows the effect of this assault erodes this command, making a successful attack that much harder.

- *Call the office every morning and tell them what time you will be arriving.* While on your way to work, be wary of what may appear to be an accident that forces you to stop your vehicle. If you see such an accident and can avoid the scene, do so. Whatever you do, don't stop to help anyone. If compelled to help, use available communications. Try to use major roads or streets at all times when driving to and from work. If you have a driver, use prearranged signals between you and the driver to indicate when there are problems.

CAR BOMBS

Although not all bombings are carried out with car bombs, it is inevitable that terrorists will make frequent use of their most popular weapon—explosives—in conjunction with one of the most easily accessible devices: the automobile.

The term "car bomb" is used to describe a number of devices. It can refer to a bomb planted in a car intended to kill those people driving the car, a bomb placed in an automobile set to explode and destroy a target near the car, or an explosive device located outside and nearby a car which is intended to destroy the vehicle and its occupants.

It is impossible to be fully protected against a car bomb by the very nature of that device, but there are countermeasures which can reduce your chances of becoming a victim of a car bomb attack.

- If the vehicle is at home, it should always be parked in a garage—regardless of how soon it may again be needed. *No one* should have access to your vehicle.
- Garage doors should be closed except for those times when a vehicle is actually entering or leaving the building. When closed, the garage doors must be locked.
- A shaft-driven automatic garage door opener should be installed, as a means of both securely locking the door and as a way of reducing the vulnerability of attack when opening the door. Chain-driven automatic doors can be forced and should therefore not be used.
- Garage windows should be securely locked and protected by either bars or a grate.
- The garage should be protected by an alarm system, activated whenever the building is unoccupied. The alarm system should have an audible signal.
- When the vehicle is at work, it should be parked in specially designated areas protected by guards.

- Access to special parking areas should be strictly controlled, not only by security personnel but also by fencing. The parking areas should be well lit and patrolled.
- Obviously you use your car for more than commuting to and from the office. When out shopping, or out for a night on the town, try not to park where the car will be unattended. If you have a driver, instruct him to stay with the vehicle when it is not parked in a secure area. If you do not have a driver, you should park the automobile in a well-lit, attended parking lot or garage.
- Vehicles should be kept locked at all times. Locks should be installed on the hood and gas cap.
- Install an alarm system on the vehicle. This system must safeguard not only the doors, but the hood, trunk lid, and gas cap as well. Most automotive alarms are designed to produce an audible signal and temporarily disable the vehicle to prevent its being stolen. In this case, your chief concern is not theft but the ability to detect signs that someone has tampered with the car. Although a standard alarm system is better than nothng, a silent alarm system—which is used to indicate that the passenger compartment, engine compartment, or trunk has been entered—is much more useful for warning that a bomb may have been placed in the car. In most terrorist environments businesspeople do not have to be concerned about bombs, but an alarm system on the car is a practical idea. The best type of system to get is one that trips an indicator located in an inconspicuous area of the auto body where it may be overlooked by a terrorist but is obvious to a person checking the vehicle prior to use.

If, for any reason, you are concerned about the possibility of a bomb being placed in your car, you should take the following precautions:

- Whenever the automobile has been out of the owner's control, it should be inspected before it is entered, engine started, or the vehicle driven.
- The inspection should begin with a survey of the area surrounding the vehicle, with particular attention directed to bits of wire or string which might indicate that it has been tampered with while unattended.
- Look for clumps of dried dirt which may have been dislodged from beneath the vehicle.
- After checking the area around the vehicle, an inspection of its exterior should be carried out without moving or touching it. During the exterior inspection, look for misalignment of doors, windows, the trunk lid, and the hood. In making the external inspection, particular attention

should be given to the engine compartment, gasoline tank, exhaust and muffler system, wheel wells, and wheel covers.

- The vehicle's interior should then be checked by looking in through the windows to see if anything appears to be amiss.
- After examining the interior from the outside, the car should be entered on the passenger's side and the interior checked. Special attention should be given to the area underneath the dashboard and the seats.
- Hanging wiring, loosened wiring harnesses, and unidentified wires leading to the ignition and hood release may indicate the presence of a car bomb.
- If, at any time during the inspection, something does not appear to be normal, the inspection should be halted and experts called in to conduct an examination.

COMMUNICATIONS

Your vehicle and all vehicles used by your company's upper management should be radio equipped. Be advised that some countries do not allow two-way-radio use by civilians. Check the local regulations. If you can legally use a radio, select one that is durable and can be serviced in the country you are in, not sent abroad for service. Although in many areas of the world an unusual antenna protruding from a passenger car is not an uncommon sight, in a terrorist environment it's a good idea to keep the antenna hidden as long as reception and transmission are not affected.

A variety of antenna configurations are commercially available. Some can be hidden in sideview mirrors, others wrap around the interior of the roof, or some can use a commercial AM radio antenna. No matter the configuration, the prime consideration must be reception and transmission signal quality. The quality of the signal must not be downgraded simply for the sake of a low-profile appearance.

The location of the microphone inside the vehicle is important. If you sit in the rear seat, you should have a microphone within your reach. If something happens to the driver, you'll want to be able to use the mike to call for help.

If local laws allow, your car should be equipped with a good electronic siren and public address system with a hidden grill speaker. It also should be equipped with a series of sensitive exterior microphones which allow you to talk to people outside the car and hear people outside the car without having to roll down the windows. This can be a tremendous advantage when driving up to a roadblock or any other situation where you don't want to get out of the car.

Your car should be equipped with a trauma kit and not just a first aid kit. Consult a physician to know what to include in the kit. A trauma kit is not your run-of-the-mill first aid kit, but a kit designed for the treatment of gunshots, burns, stab wounds, and heart failure. The trauma kit contains the materials necessary to stop bleeding, maintain a clear air passage, and keep a person alive until a hospital can be reached. The kit should be kept inside the vehicle's passenger compartment for ready use, not stored in the trunk. All hands, including yourself, should have emergency medical treatment training and know-how to stop bleeding, control shock, and perform cardiopulmonary resuscitation (CPR). Basic first aid skills, along with the ability to perform CPR, are things that everyone should know how to do.

CAR OPTIONS

It is impossible to overstress the importance of a secure vehicle in your personal security program. You must have a safe and secure vehicle. Let's look at the options you'll want:

- If you are buying a car, these are some good factory-installed options that may prove useful. The most important is any optional higher-powered engine. If available, get the biggest engine that car will take.
- Below-surface or smooth door locks resist attempts from thieves trying to enter the car easily.
- Trunk and hood locks make it harder to find a place to put a bomb or disable the car.
- Locking gas caps prevent sabotage.
- Dual side mirrors with remote controls help you check for surveillance.
- Power steering, for quick and easy handling.
- Air-conditioning, so there is no need to roll down windows.
- Power disc brakes, for emergency stops.
- Steel-belted radial tires, for better handling.
- Rear window defroster, if applicable, so you can use your inside mirror to watch for surveillance.
- Heavy-duty suspension, for better handling.
- Under-hood-mounted siren or foghorn to surprise your attackers.

The following car options are worth considering if your threat situation is significant:

- High-intensity headlights, to see farther at night and give you added time to react.

- High-intensity backup lights. These give you better vision to the rear at night if you have to back up fast.
- A communication system but no external antenna other than AM/FM radio, to report detours or delays, and to call for help.
- Grill-mounted police or flasher lights for surprise and escape. When these are mounted within the grill, they cannot be seen when they're turned off.
- Remote radio-controlled ignition starting system so the car can be started and warmed up without anyone in it. This is by far the safest way to check for bombs wired to the ignition.

DRIVING IN A TERRORIST ENVIRONMENT

When you leave your car in a parking garage and are required to leave a key with an attendant, leave only the ignition key. If you leave all your keys with the attendant he can easily make duplicates of all the important keys in your life while you are out, since your house and office keys are probably on the same chain. The attendant can even take his time making those duplicates because he probably asked you (and you probably told him) when you'd be back to pick up the car.

While he's deciding which of your keys to have copied, he can also look through the glove compartment for the car's registration or other documents that might reveal your address. Now our intrepid parking lot attendant has the key to your house, your address, and other documents which might have yielded other useful information. If you're going to be this accommodating to a complete stranger you may as well go ahead and join a terrorist group to save them the trouble of kidnapping you.

When you leave your car with the attendant, note the odometer mileage on your claim check. Do it so the attendant can see what you're doing. This may prevent either him or his brother attendants from taking your car for a joy ride.

Don't leave anything of value in the car, especially where it is visible.

Place everything in the trunk, lock it, and take the trunk key with you. Don't forget your claim check!

If you are parking in a shopping center or a public parking garage, always study the situation before getting out of your car. If anything looks suspicious, drive off and find another parking spot or come back later. Always try to park in a well-lit area.

If you are ambushed and your car suffers damage, keep in mind that a car can take a great deal of abuse. If a radiator is punctured (by being shot full of holes, for example), a car can continue for another eight to ten miles before dying.

Having four flat tires makes driving rough, but it may be a lot rougher to stop and deal with the gents that just shot those tires flat. Flats don't force you to stop. Don't worry about the tires (or what's left of them) and don't worry about the wheel rims. Just GO!

If your fan belt is broken the car will overheat, but it can still be driven.

Keep your gas tank at least half full at all times. You may need it.

Part of keeping a low profile is driving a low-profile car. Corporate logos on cars can draw unwanted interest from "the wrong sort of people."

If you do drive your own personal car, select a vehicle common to the locale (make, model, color) rather than an exotic custom design. Discourage personalized license plates or distinctive car accessories. And sporting a U.S. flag from the fender, diplomat-style, may not be interpreted as commendable patriotism by the locals.

Each time you get into the car make sure you look for someone lying on the back seat or on the floor. If there is a suspicious package inside the car, or just something that doesn't look familiar or right, DON'T GET IN THE CAR! Check the area for other parked cars with passengers. Make sure that when you leave no one leaves right behind you. Watch for cars that pull out when you do and follow you to your destination.

USING CHAUFFEURS

Chauffeurs are common in many parts of the world. If you are fortunate enough to have your own driver, here are some simple rules to follow:

- If, instead of your regular driver, an unfamiliar driver shows up some morning to take you to work, establish the validity of that driver before setting foot in the car. If that means going back into the house and calling the company or limo operator, do it. As a child, your mom taught you not to get into a car with any stranger. It's especially true today.
- Have a prearranged signal, to be used by your chauffeur if he's under duress, warning you against entering the car. This could be a visual signal given as you're walking toward the car if all is not right. It could also be a spoken signal, such as that given over the phone in a situation in which terrorists have nabbed your chauffeur and want him to call you to arrange to pick you up. An innocuous code phrase you both understand, such as "Your Aunt Ella called the house," when both of you know you have no Aunt Ella, could serve.
- The driver should always open the door for you. Do not enter the car unless he opens the door, from the outside.

- If your driver is not there when you arrive at the car, don't get into the car anyway. A good way to prevent this is not to go out to the car until your driver has reported to you.

Once you have arrived at your destination:

- Always lock the car.
- Do not leave your car on the street overnight.
- Be alert in underground garages. Do not exit the car without checking the area for suspicious individuals. If you see any, drive out of there.

ARMORED VEHICLES: AN INTRODUCTION

Never lose sight of the fact that the primary mission of any vehicle is not only to protect the occupant but to transport you from point A to point B. The objective of the vehicle is to get you wherever you are going in comfort, efficiency, and with a certain amount of performance.

Armored cars are very common in many countries. Know that by the term "armored cars" we do not mean those gun-metal-gray rolling safe-deposit boxes you see on city streets transporting large amounts of currency between banks. The armored vehicles we are talking about are cars, limousines in some cases, that have been fitted with armor plate in a way that takes nothing away from the car's good looks. One of the best things about these types of armored vehicles is that they don't look armored.

If your security department feels you need an armored vehicle, don't fight them. Like prescribed medicine, an armored car is for your own good. When selecting an armored car, the type of vehicle you select must fit the environment of the country. You do not want an armored Cadillac if it will be the only Cadillac in the country. In Italy, a Fiat is less conspicuous than a Cadillac; in Germany a Mercedes; in France a Peugeot. Maintain a low profile. Pick a color common to that model car. Stay away from flashy colors like bright red or yellow. Bear in mind that in many societies a black sedan represents official status or wealth.

See if you can find out what the local government's favored vehicle is. The imprudent choice of a car that just happens to be that of unpopular officials could make you a target for terrorists who might not take the time to follow you around and get to know you, but instead just try to kill you on the spot for the sake of what they think you represent.

If the company decides to buy an armored vehicle for you, you should have a lot to say about the vehicle. You will be the person sitting in it if someone takes a shot at it.

There are just a handful of protected vehicle manufacturers in the U.S. Great improvements have been made in the design and construction of protected vehicles, to the point where most armored cars are built in a way that the armor is barely noticeable. The benefits of armor are not the only consideration in a protective vehicle. Since armor can add a great deal of weight to the entire design, it's important to make sure the vehicle remains maneuverable. The more armor you put on a car, the more you decrease the vehicle's acceleration and maneuverability. These characteristics can be every bit as important as the armor. You don't want an armored car that is too heavy or too unwieldy to drive.

All unprotected areas must be eliminated on an armored vehicle. Glass is the most vulnerable area. If your vehicle is attacked, be assured that the terrorists will concentrate on the glass. Bear in mind that bullet-resistant glass means just that; it is bullet *resistant,* not bullet *proof.* Concentrated attacks on these surfaces will eventually result in bullets penetrating the glass and hitting the car's occupants.

Among the great advances made in the manufacturing of protected vehicles are cars equipped with Safoam, a plastic sponge like material, which can be installed inside the gas tank and does much to reduce the danger of explosion or ignition of the gas tank if it is pierced by gunfire. Safoam can serve another practical safety purpose. If your vehicle is rear-ended, Safoam can prevent the gas tank from exploding.

Your armored vehicle also needs extra-heavy-duty springs and shock absorbers; other options can include things like concealed gun ports; armored battery boxes; high-intensity pursuit-deterrent lights; vehicle tracking systems; smoke and tear gas dispensing systems. Another device that should be on all cars are run-flat-tire devices. These will enable the vehicle to leave the scene of an attack at speeds up to 50 miles per hour even if all four tires are punctured.

All of these devices cost money, and a decision has to be made as to whether the risk justifies the cost.

No matter what vehicle is selected, you must demand that the vehicle be tested. Because so much weight has been added to the car, make sure it stops in a reasonable distance. For the same reason, test to see if it can accelerate on level ground or on a hill. Make sure it can climb hills without laboring. In short, make sure it can be driven in the environment it will be used in.

One reminder—possession of an armored vehicle does not translate as complete and effortless safety. Don't be lulled into a false sense of security by driving an armored vehicle. You still must take all the security precautions you would in an unarmored vehicle.

Level of protection: Armored cars can be purchased equipped with different levels of protection designed to stop certain weapons' calibers. Standardized by a system developed by Underwiter's Laboratories, the levels of protection are:

Level II—will stop a .357 magnum or 9mm handgun

Level III—same as above, but will also stop a 44-magnum, 12-gauge, .30 cal carbine.

Level IV—same as above, and will also stop heavy 30.06 military ball ammunition.

The level of armoring you need is determined by the type of weapon you want to protect against. An example is the UL's highest rating, Level IV, which protects against 30.06 cal, 24-inch barrel, soft grain slugs traveling at 2400 ft/second. But muzzle velocity for a slug from a NATO-class weapon can be as high as 2900 ft/second. So there are some commonly encountered weapons that can easily penetrate even this heavy armor.

From UL's standardized system, it appears that in order to establish the degree of protection needed, one must know the type of weapons that may be used in an attack, and the method of attack, such as firing from a distance versus close range. Weapons fired at close range carry far more power.

Kidnap teams prefer automatic weapons. Experience and many studies show that the average terrorist attack will be at close range. Combining this data, we can conclude that your level of protection should be at least Level IV or higher. At this stage, a consultation with an armorer can be of great assistance. It is the armorer who knows as much about the levels of protection needed for certain areas of the world as anyone. If he has not constructed a car for that part of the world, you shouldn't be his first customer. Instead, find someone who has. And, as always, any information he imparts to you should be backed up by your own intelligence network.

Type of vehicle: There should be ample room in the car for four people. Therefore, the car selected should:

1. Blend into the local country's vehicle environment, not be flashy, and not stand out.
2. Have four doors—comfortable and allowing ease of entrance and egress.
3. Have a power plant and gear ratio sufficient for reasonable acceleration.

Vehicle design: The armored vehicle should be designed to:

1. Absorb the attack, take repeated hits, return fire, and call for help.
2. Absorb the initial fire and break the ambush.
3. A combination of the above.

Please understand that of the above, the first capability is extremely expensive. The type of vehicle that can absorb an attack, take repeated hits, return fire, and call for help should be used against major threats such as political assassination. The cost of such a vehicle is directly proportional to how long you would like to sit there and absorb fire. Cars armored to maximum ballistic levels are very expensive. Prices can run as high as $275,000.

Any armored car you drive must be capable of absorbing the initial burst of fire from the type of weapons used by terrorists and still have the capability to get itself out of the ambush.

Armoring the vehicle: The car purchased must have the doors, windows, and rear of the passenger seats sufficiently armored to counter the maximum threat level. If grenades are a potential threat, the roof and floor must be armored. If this becomes necessary, the price of the entire vehicle rises dramatically. You will also want to armor the car's fire wall. A terrorist standing at the 10 or 2 o'clock position in front of the car could fire at a fender and the round could penetrate the fender, go through the fire wall, and hit one of the passengers. Armoring the battery, radiator, and fuel tank(s) requires discussions with the armorer. A foam-filled fuel tank is well worth the investment. A round hitting the gas tank can be disastrous.

Armoring the radiator can sometimes create more problems than it solves. If your radiator is punctured you can still drive as far as eight miles at highway speeds, less at slower speed, and armoring the radiator can cause overheating problems in some car designs. Some armored cars include louvered radiator guards that can be opened for normal driving and closed for protection in the event of an attack. The problem with this design is that few attacks carry much warning, and therefore would give the driver very little time to button up the radiator before the slugs started hitting.

Additional equipment: This should be considered on all armored cars. Some of the equipment may not be pertinent to your particular situation, but it's worthwhile to consider it anyway. Take a look at the following options:

1. *Steering*—a heavy-duty steering pump or a power steering cooler. If quick steering maneuvers are needed, the steering fluid can foam and cause loss of steering.
2. *Communications*—the occupants of the vehicle should be able to communicate with someone at the place of business. The higher the threat, the more elaborate the communications. What is as important as the

hardware purchased is how it is used, and procedures on how and when to use the equipment must be made clear to all involved.

3. *Mirrors*—Good rear- and side-view mirrors should be installed on the car so the driver does not have to move his head to look to the rear. There must be side mirrors on both sides of the car, and they should be remote-control mirrors. The external mirrors should also be flat mirrors to give you better, more accurate vision to the rear.
4. *Covert antennas*—if you have a communication system in your car, you don't want everyone to know about it. There are specially designed antennas that fit into the skin of the car and cannot be detected.
5. *Horn*—a loud horn or speaker system can alert passersby and may scare your attackers into calling off the assault.
6. *Public address system*—the driver and occupants should be able to talk to someone outside the car without opening the doors. Also, the driver does not have to drive up to someone to talk to him; the car can stay a safe distance away from strangers and communicate with the windows still rolled up. Such a system can also be used to hear what the people outside the car are talking about.
7. *Tires*—your car must have run-flat devices on them. The run-flat device allows you to drive the car if the tire goes flat. There are times when you may get a flat in the wrong section of town and don't want to get out and change it. There may be times when terrorists have shot your tires full of holes and you do not want to follow suit.
8. *Optics*—be sure there is little or no distortion through the windshield. When purchasing a car, the ambient temperature of the environment is important. If the window is not designed properly, temperature can distort the view quite badly.
9. *Engine protection*—depends on where the executive lives. In an urban environment, you would probably need to get just a short distance away from an ambush in order to be safe. An engine will keep running without water for a sufficient amount of time to get clear of the urban ambush. But in rural settings, where the driver may need to drive 50 plus miles before he can stop, it may be smart to armor the radiator.
10. *Chassis*—if the car is sufficiently armored there should be some modifications made to the chassis. If the gas tank is protected by armor, the rear frame definitely needs modification.
11. *Tool kit*—there should be a tool kit in the trunk of the car. A hidden tool kit can be used to free oneself from the inside of the trunk in the event that abductors put someone there.
12. *Testing*—when driving a car make sure it can stop in some reasonable distance. Stop on an uphill grade and test the car's ability to accelerate up the hill. If the car struggles to make it up the hill, it's not worth the price!

HOW TO BUY AN ARMORED CAR

Comparison shop before you buy any armored car. Get bids from different manufacturers. Ask for recommendations. Don't accept the standard reply "recommendations are kept confidential" as an answer. If he can't trust you, who can he trust? Determine what the percentage increase in weight is for an armored vehicle as opposed to its unarmored counterpart. If the weight increase approaches 25 percent, vehicle performance will be severely limited. Don't take the armorer's word for it—weigh the car yourself.

Be very fussy about the optics. Ask how much distortion will occur in the windshield. There should be little or none. Ask to see the dealer's quality-control procedures. The ratio of quality-control workers to the assemblers is important. The more quality-control workers the better.

The life blood of an armored vehicle is the composition of the glass and the armor material of the vehicle. Material type is important: Kevlar®, Ceramic, and Dual hard steel are some of the materials used, and every armored car builder can show you pieces of armor that have been shot at and can discuss for hours the disadvantages of the other materials. The best way to address this problem is to test it for yourself.

Your security department might object, claiming it is too complicated or too time consuming. If they tell you that, ponder for a moment, perhaps out loud and within earshot of your security team, just how much you are spending for this car, how much you or your company is spending on the security department, whether the latter amount is truly worth it, given its reluctance to help in testing materials, and you should soon find yourself benefitting from the professional advice and expertise of your security staff.

Levity aside, you've got to test both the armor and the vehicle. If you wait for terrorists and/or kidnappers to do it, it may be too late. The following is an outline of how to test the bullet-resistant capabilities of the vehicle.

Testing: Your company must be allowed complete access to the manufacturer and retain the right to select a portion of the vehicle for ballistic testing at any stage of production. For example, a door may be removed and shot at within the contract parameters. Any testing will be done in the presence of qualified representatives of both the manufacturer and the client. Weapons and ammunition used in the test will meet prior contract specifications agreed to in writing. Testing procedures will be agreed upon in advance by both parties. The manufacturer will replace the tested part for the cost of materials and labor.

You can understand that such testing will cause a production delay. The fact that you have the option to test the vehicle in this manner may be sufficient. At random, the vehicle must be checked periodically to insure that the entire passenger compartment is protected and that there are not "gaps" at the doors, back seat, windows, and fire wall.

The above measures may seem a bit harsh, but keep in mind what you are dealing with. You are investing a great deal of money and (hopefully) protecting lives, using someone else's product. You must have the right to test the quality of the product. The very worst thing that can happen is that you create a false sense of security. That false sense of security can and will kill you just as dead as terrorist bullets.

Vehicle performance: It is not difficult to measure the vehicle's performance—you can get a good idea of what it will be by simply measuring the weight of the vehicle before and after armoring. Therefore:

A. The vehicle weight should be measured before and after armoring. The weight increase should be no more than contracted for.

B. Weight distribution front to rear should be checked. Much depends on the original weight distribution; no more than 55 percent of the weight should be on the front of the car. This is a fairly accurate number for rear-wheel-driven, front-engine cars.

C. Make sure the tires can take the additional weight. Look on the side of the tire and it will indicate the maximum weight acceptable at a certain tire pressure. If you weigh the car you will receive the weight on the front tires. Also remember to leave some margin for weight transfer to the front under braking.

Quality control: The only way to measure the quality control of the vehicle is, with the manufacturer's permission, to discreetly contact others who have had cars manufactured for them and to solicit their impressions of the product. There are a number of questions that should be asked, such as:

A. Did the manufacturer meet the delivery schedule? If not, why?

B. Was the car weighed prior to and after modification at a licensed weigh station? Did it meet the prescribed added weight requirements?

C. Does the vehicle feel top heavy or off balance, particularly at high speeds?

D. Have there been any problems with:—

- **a.** transparency distortion and glass/acrylic delamination; if so, how were they solved?
- **b.** vehicle ride. How smooth was it? Does the car drift or skid easily?
- **c.** rattles or excessive noises in the car? These may indicate improperly installed armor which has been jarred loose by road impact.
- **d.** doors hanging and operating improperly? Improper hinging is a major weakness of a less-than-expert company.

e vehicle performance, sluggishness, poor acceleration or maneuverability, engine overheating, loss of power on inclines, excessive tire wear or poor braking.

Company background: You need some answers on the firm building your car. Among them are:

A. Length of time the firm has been in business
B. Technical and manufacturing capabilities
C. Warranty policy
D. Reputation in the automotive field

9

Transitory Protection: Traveling by Land and Air

INTRODUCTION

Protection while traveling by means other than private car is an important part of personal security. International travel and travel by public transportation within foreign nations can be a harrowing experience.

Here we're focusing on two basic scenarios. The first centers on problems associated with traveling to an unfamiliar country, and the other on moving about the country you are living and working in. Both create special problems for you and your family.

At one time, traveling to another country by commercial airline meant a pleasant and generally interesting journey, and little more. It was nothing to be concerned about. But terrorist attacks, skyjackings, and prolonged hostage situations have made air travel much more worrisome for all travelers. Despite these hazards, air travel is still the safest form of transportation. Unfortunately, terrorist incidents do happen, and when they do, Americans are singled out as targets. Air travel is an undeniable problem, but one that a lone individual can't do much about. You can be far more effective protecting yourself and your family on the home front in the overseas country in which you are stationed. You probably spend far more time traveling within your own city than flying internationally, so the question of in-country travel merits close attention.

INTRACITY SAFETY

Traveling within the confines of any major city in the world presents a security problem for you and your family, not because of terrorism, but due to plain old street crime. Sadly, plain old street crime can be just as deadly and hurt just as much as terrorism. Whether you are visiting a country and touring the city streets, or are in your country of residence traveling the streets you need to use in your everyday business, here are some simple steps to stay out of trouble. Some of these are simple rules that have not changed in centuries, such as don't go out alone after dark unless absolutely necessary. If you do feel compelled to go out, try not to go alone. The more numerous your group, the safer you are.

If you have decided to play Russian roulette and go out alone after dark, or if there just is no other way, walk on the curb side of the sidewalk. Stay away from dark storefronts and blind corners. These are great places for thieves to hide and get ready to make whatever belongs to you belong to them.

If you decide to stroll around the city, stick to well-lit, populated streets and districts. Use some common sense and stay away from the city's "combat zone," that is places where there are prostitutes, strip joints, or sex shows. These places invite trouble. If you have an inclination to visit such places and get into trouble while you're there, it may be awfully tough explaining to your spouse and friends just what you were doing there. What's more, if you do get into trouble in a sleazy trouble spot, you'll not get much sympathy from the local police. They'll probably feel that you should have known it was not a nice place to be in in the first place, and that since you were there of your own free will, you accepted the consequences.

If you don't know where these places are located and are afraid of blundering into them by accident, ask local taxi drivers, bartenders, hotel bell captains, and others who keep their ears to the ground. These people know the pattern of local crime. Pick their brains about dangerous places and times in their city. In almost every situation described in this book, even international terrorist activities, these people can be indispensable intelligence sources. A well-connected bartender in some Third World situations probably knows more about the day-to-day scene in your city than the local CIA station chief.

In almost every city there are places that are safe at 5 P.M. and dangerous just a few hours later. Even if you are familiar with an area from previous trips, don't assume that the place is still the same. A location that three years ago may have been a nice place to visit may now be deadly.

This doesn't mean when you visit a city you have to lock yourself up in your room and watch TV all night. You can wander the streets freely as long as you know where you are going. There are other things that you can do to lower your level of risk on the street day and night. The most important is to make an effort to look unimportant, blend in with the local population. Dress down. A "foreign," out-of-town look with expensive clothing and flashy accessories—briefcases, cameras, watches, cufflinks, tie clasps, or rings and other jewelry—will set you up as someone with money. The local street thugs are attracted to people who look like they have money. After all, why rob someone who is poor? As you walk down the street, put yourself in the hoodlums' shoes. See how many "foreigners" you spot that would be easy targets.

No matter where you are, take some basic precautions. Keep some space between you and other people on the sidewalk. If you see someone walking towards you who looks suspicious, or just plain strange, give them plenty of room. When you're walking the streets, be wary of people who want to talk to you. A common thieves' trick is to ask for the time, and as you are looking at your watch, grab it off your wrist and run away. Or an accomplice takes your wallet or purse while you are occupied.

Don't respond to approaches from street people, especially street vendors selling everything from precious stones to jewelry for bargain prices. The time to stop these people is when they first approach you. Don't get pushy about it; be polite but firm. Don't stop to talk to them; don't even slow down. Every country has some product that it is noted for, and can be purchased inexpensively. If you want to purchase that product, go to a reputable store and buy it. Ask a business associate where you can get a reasonable deal at a reasonable price. Don't buy from a street peddler or go with some taxi driver who knows where he can get it really cheap. You may wind up with no product and lose all your money in the process.

While we're on the subject of shopping, be careful about your credit cards. Sales clerks have been known to validate two charge slips, then trace your signature on the "extra" slip, and fill it in later with extra items. Try to keep an eye on your card. This is usually impossible to do in restaurants, so the best thing to do is to eat only at good restaurants with good reputations. *Take the credit-slip carbons with you and destroy them!* Criminals can make an imprint from the carbons and have free use of your credit card for quite a while.

A relatively new form of convenience has created a new form of crime, robberies at 24-hour-a-day automatic bank tellers. You're unlikely to encounter these overseas, but use of them is increasing in the large cities of some of the modern nations of Europe and South America. Be very careful around these devices, especially at night. If you feel there is a problem developing, purposely enter an incorrect code for your card. Do it three times and the machine will take your card. That way you will not have to surrender your card and cash to someone who is about to rob you. The bank will return your card on the next business day, most likely at the branch nearest the machine that swallowed the card.

Although it is not recommended that you talk to strangers, there will be traveling situations that arise in which you cannot avoid having a conversation with people such as airline ticket agents, hotel staff, waiters, cab drivers, and a broad spectrum of people you come into everyday contact with while traveling. You have to face the fact that there will be times when your contacts with these people may not be pleasant. There's nothing like having a hotel clerk tell you that the reservation you made weeks ago is not there. There's a perverse logic that says this always happens after long, grueling, unpleasant airline flights, not after pleasant trips. In such a case, be firm but polite. Don't make enemies.

Don't be an overbearing, hard-to-please "Ugly American." Try to be a friendly, good tipper who develops friends wherever you go. A little bit of attention, kindness, and concern in dealings with others while traveling, especially those lower on the economic scale, can pay life-saving dividends in the event of serious problems.

Nothing creates future revolutionaries quite like pushy, ill-mannered, impatient American visitors who treat the locals like some species of thick-skulled beast. Be nice. It pays.

Away from home base, you may have currency problems. Always use travelers' checks; never carry large amounts of cash with you. Make sure you are carrying the types of credit cards that allow you to get cash advances against your credit limits. These can make your life a lot easier in the event of any problems. If you do go to a bank or money exchange to change currency, never flash your money around. This can be most hazardous to your health. Criminals hang out around banks and money exchanges checking out the people as they leave. Counting your money as you're leaving is like waving a flag and asking to be robbed.

Don't laugh. All this seems like good basic common sense, but the next time you're near a bank or money exchange, count how many people are walking out while counting their money. Count your money at the window and then put it away before going out into the street.

PUBLIC TRANSPORTATION

If at all possible, try to avoid public transportation. Although very safe in many countries, why take the chance?

Taxis are the most common form of city transportation. Cabs are often the safest way of getting around in most major cities. Most cab drivers know the bad sections of town and, if asked, will tell you what parts of town to stay out of.

Here are some precautions you should consider before taking a taxi. If you feel your threat level is high, and you are in your own country, never take the first cab in line at a taxi stand. Terrorists are your concern in this situation. A terrorist could easily masquerade as the taxi driver and wait for you to come out of a building or meeting and walk into their trap. But if you are on vacation or visiting a foreign country, taking the first cab is not a major problem.

Once again, it is much safer to travel in pairs or groups by taxi. Pick a cab that's empty—except for the driver. This may seem obvious, but in some countries it's a common practice to have more than one passenger in a car, so don't get in the car unless you are the only passenger. Then ask the driver not to pick up any other passengers. If the cab is unmetered, agree on a price before beginning the trip. This will avoid a hassle at your destination. You may add to your security in a cab by sitting up front beside the driver, rather than riding in the rear seat. Many countries have gypsy cabs that roam the street. Find out from the embassy or a security source how to recognize a gypsy cab and avoid them. Gypsy cabs are not licensed cabs and could very well drive you to a destination other than the one you requested and at that point relieve you of your belongings and leave you there.

If you are eligible for a company or rented car, try to arrange for a nondescript vehicle that blends in well with other cars on the street. Prestige license plates, fancy paint schemes, company logos, or anything else that sets your car apart should be avoided.

Of course, when parking your car in a lot or on the street, lock it and carry the keys with you. In many parts of the world, most notably in Latin America, you will find enterprising and persistent street children who will offer to watch your car for a small fee. Hiring these kinds to "watch over your car" can be a very good idea, because it should keep them from stealing or damaging the car. These new friends are inexpensive and they usually respond well to a smile and a few kind words. Remember to pay them only when you retrieve your car. If you pay them in advance, they will probably disappear as soon as you leave, running off to find other sources of cash and leaving your car unattended.

As you would in any overseas situation in which it pays to be wary, deliberately vary your mode of transportation and travel routes and times. Most big city hotels have several entrances and exits. Use them all. Devising a complicated, unpredictable pattern makes you a very difficult target for terrorists or others who might consider kidnapping you, mugging you or just generally being unpleasant.

Maintain a low profile while traveling, with unpredictable patterns of movement, just as you would at home.

If you are mainly traveling within the country in which you have been stationed, you'll find that maintaining a high level of personal security is much easier than when you are traveling in countries strange to you. As a seasoned veteran of that country, you already know the areas you should stay out of, so most of the battle has been won. Stay unpredictable and you'll probably be OK. Discipline yourself to do that. Don't try to convince yourself that it will be OK to visit a place you know that you should not be visiting.

CORPORATE AVIATION

Terrorism is a new threat to corporate aviation. Although this may not be your responsibility, some basic information may be helpful to you in the event you travel in the company jet. If you are responsible for corporate aircraft, you should be aware of some potential problems.

Unlike flying in the U.S., where air crews have access to information about the airports they use, very little can be found out easily about security conditions at overseas aviation sites. Nevertheless, methods do exist within the security communities by which air crews can get comprehensive information about existing security arrangements at various international airports. Private flight planning services specializing in overseas operations provide excellent professional service. Your corporate flight department manager should know of several. If not, contact the National Business Aircraft Association, One Farragut Square South, Washington, DC 20006 (202) 783-9000. This organization can do much to put you in touch with specialists in overseas corporate aircraft operation.

The U.S. government maintains several briefing services which can provide intelligence about the security conditions prevalent in many countries of the world. Some of this information will bear directly on the major airports of the countries concerned. In this regard, three principal agencies have such briefing capabilities: the U.S. Department of State, U.S. Department of Commerce, and the Civil Aviation Security Service of the FAA.

For obvious reasons, these agencies prefer to render face-to-face briefings in Washington, D.C. They will give the briefings to those in the corporate community requiring such services and assistance. This face-to-face meeting is required to maintain confidentiality and to minimize the threat of countertactics by the "bad guys." Perhaps after the initial contact is made, some supplemental telephone briefings may be given.

Although terrorist actions against corporate aviation have been minimal, the following suggestions are simple and do not require much attention. As always, you want to maintain a low profile in your company plane. Do this by removing any company logos from the aircraft's exterior. Air crews must keep flight schedules secret. Operate the aircraft quietly but efficiently, making as little information public as possible.

Special self-destruct labels should be used to cover an aircraft's entrances, ports, and other vital parts while the airplane is parked overnight away from home base. This needs to be done even if you hire a security guard to watch the aircraft. Several such seals are available on the market and many corporate operators carry them to detect and prevent molestation of the airplane, along with reducing the potential for bomb threats and unauthorized access to the aircraft. Sometimes these labels are used in lieu of the special luminous paints long used for similar detection purposes.

The pilots need to perform a good preflight security inspection, for example, check oil caps for tightness; check the engine and baggage compartments for sabotage and suspicious devices, wires, or other unknown objects. On jet aircraft, they should check between the engine's intake vanes for foreign materials; check landing gear wheels to make sure the cotter pins have not been tampered with or removed; check tires for punctures and tampering; and check flaps, ailerons, elevators, and rudder hinge assemblies for evidence of tampering.

Carefully control access to the aircraft at all times. Consider placing carry-on hand baggage in the cargo compartment. Identify every piece of luggage with the particular passenger who owns it and make sure that person is indeed on board. Know precisely who your fellow passengers are. No hitchhikers—ever! Require identification for all persons. Carry-on baggage needs to be searched and, if necessary, the passengers need to be searched. Boarding should be denied to anyone who refuses to be searched. If you are bringing friends along on the flight warn them in advance that they will be required to identify themselves and that they will have their carry-on luggage searched. Your cooperation makes this program much easier to implement.

Standard policy for many corporate flight departments is that flight crews visit and inspect their aircraft each day when on an extended stay in another country, whether or not their aircraft is guarded by a contract guard. Some companies even go as far as to require their crews to stay close by the aircraft at all times. This is why many of the airport-based firms servicing business jets include dormitory/motel-type sleeping rooms at their facilities.

Upon arrival at the destination airport, flight crews should develop the habit of scanning the airport operations area where their aircraft is being directed to park. It's important that the flight crew evaluate the risk of surprise intruders who could easily emerge from a structure or parked cars near the aircraft's parking area, particularly at overseas airports. If possible, avoid parking aircraft in the vicinity of gates or low fences designed solely as barriers for crowd control rather than for security protection.

The crew should monitor any servicing of the plane, including refueling, maintenance, and the stocking of food. Food and drink should be secured only from reputable aircraft caterers. The business aviation facility you patronize will be able to order reliable catering for the aircraft.

Park the plane in a designated secure area at the airport. If the plane must be left overnight, be sure it is parked in a well-lit area.

Set up a means of round-the-clock communication that can be used between the crew at the airport and any passengers should the crew receive a threat, or if local conditions call for an unscheduled departure.

Typically, corporate aircraft are locked but left unguarded while they are on the ground. Anti-intrusion alarm devices can alert authorities should someone attempt to tamper with the plane. Alarm systems that use electromagnetic switches are usually best because they are not affected by the high vibration and noise levels common to an airport environment. They also do not interfere with radio frequencies used by the control tower and the aircraft.

However, local law enforcement authorities, airport security staff, and others who regularly patrol the airport must be briefed on the alarm system. They also must know how to reach your crewmembers should an emergency occur.

A professional approach to corporate aviation security is absolutely necessary to develop an adequate, cost-effective program. The following steps are recommended:

- Assess the threat to the individual corporate aviation activities. The threat can vary greatly depending on the type of aircraft involved, locations visited, and the company's involvement in international and/or controversial issues.
- Conduct an in-depth security survey of the corporation's aviation activities, equipment, and facilities.
- Prepare a comprehensive corporate aviation security program that will correct deficiencies and satisfy needs noted in the survey. There are excellent organizations that offer training in corporate aircraft security. These organizations are listed at the end of this book.

PERSONAL SECURITY GUIDELINES

Here are some general personal security guidelines that you should follow when traveling:

- You will need to exchange money when traveling. Try to obtain any necessary foreign currency and/or travelers' checks before you leave the airport. You may get a better rate someplace else, but doing this at the airport is far safer than exchanging your money at the corner moneychangers.
- Outside the U.S., public phones are strange devices. Learn to operate the local public telephones and carry the coins necessary to operate them with you at all times. Also carry a card listing local emergency

telephone numbers, such as police, fire, paramedic, or hospital, along with your local company representatives, and of course, have the U.S. embassy's phone number with you at all times. In fact, the first thing you should do when you arrive in a country is locate the American embassy. Know how to get there, and get and keep its phone number.

- *Never leave your passport, credit cards, or other valuables unattended in your hotel room.* Never, never, never. Ever. Guard your passport as though it is worth a fortune, because on the black market your passport *is* worth a fortune.
- Always have extra passport photos available for immediate use.
- Carry a photocopy of your passport that shows its number, date, and place of issuance. Keep this copy, along with an extra passport photograph and pertinent visa material, separately from the original.
- Keep a photocopy of your current passport and credit cards at home and a duplicate at work.

HOTEL SECURITY AND SAFETY

Avoid ground-floor rooms to reduce the possibility of an unauthorized and otherwise unwelcome and probably generally unpleasant person entering your room through a ground-floor window.

When you are in your room, keep your door locked at all times. Place the "do not disturb" sign on your door.

Place a chair or piece of luggage by the door to alert you to trespassers.

You may wish to carry your own slip lock or portable alarm.

Report defective door or window locks to hotel management immediately.

Be suspicious of unusual calls to your room. Do not admit visitors without fully identifying them first. When you do leave your room, make sure the door is locked. If it is evening and the room is made up, leave the "do not disturb" sign on the door, and leave the TV or radio on. Take your room key with you. *Do not leave it at the desk!* You want to create the impression that you are in the room.

Lock all valuables and extra cash in a hotel safe deposit box.

Be careful what you leave lying around the room. If you walked into an occupied hotel room, you could probably tell who the occupant was, their occupation, and their relative importance. Items found in hotel rooms can mark you as a potential target; items such as luggage tags, clothing, business documents, correspondence, money exchange receipts, and airline tickets can well document a person's relative worth as well as whether it would be worthwhile to stick around and rob him or her. Or worse.

Everyone realizes an expensive set of matched luggage makes a statement, as does a hand-tooled leather briefcase. But a common suitcase that sports a tag identifying its owner as the senior vice-president of a major multinational corporation says much more. Business cards or travel itineraries carelessly left lying about may compromise personal security.

Marighella's *Mini-Manual of the Urban Guerilla* preaches the importance of an intelligence network to the urban guerrilla. An active intelligence network involving waiters, chambermaids, bellhops, reservation clerks, and switchboard operators flourishes in many countries. This network functions primarily for valid reasons. Hotel security people like to detect and frustrate those who might pose a threat to other guests (say, jewel thieves) or who might attempt to skip out without paying their bills.

Due to the existence of such an intelligence network, a lot of extraneous data and gossip about guests is routinely collected and passed on. It's easy for a terrorist group to tap into this rich source of information.

To get a jump on this kind of activity, business travelers must think like terrorists. Put yourself in a terrorist mindset, take a good long look at yourself in the mirror, and ask, "Would I go after the person reflected here?" If the honest answer is "yes," consider why. Then do whatever needs to be done to change the answer to "no."

Some spectacular hotel fires have killed people and grabbed a lot of headlines in recent years. Be prepared for the worst. When you first get to the hotel room, locate the nearest fire exists. Count the doorways and other hallway features between your room and the emergency exits. In the event of a fire, you may not be able to see where you are going, so you may have to rely on your sense of touch. While you're exploring, find the nearest fire alarm.

Examine the layout of your room. You may have to remain there if smoke in the corridor cuts off your escape. Make sure you know how to open the windows in your room. Make sure they work.

If you're sure there's a fire, do the following:

- Telephone the fire department, then notify the hotel operator (be sure to give your room number).
- Check your room door; if the door is hot, *do not open it!*
- If you leave your room, take your room key and go to the nearest fire exit. If the fire exit is blocked, you will need your key to get back in your room. Stay beneath any smoke that may be in the corridor, crawling on the floor if necessary. *Do not use the elevators!*
- Fire exits generally lead to stairwells. Enter the stairwell and close the door behind you. If possible, go downstairs and out of the building. If you cannot go down and out, pause for a moment and think clearly. Use your best judgment—either go to the roof or go back to your room.

- If you stay in your room, turn off the air conditioner, fill the bathtub with water, wet the sheets or towels and stuff them completely around the door, and block vents that are emitting smoke. Open the window only if you are sure fresh air is available. Again, call the fire department—give them your room number and tell them you are staying in your room. Hold wet towels over your nose and mouth. If the doors and walls are hot, douse them with water using an ice bucket.
- *Stay calm; don't quit.*

BASIC TRANSPORTATION

- Restrict knowledge of your itinerary to office and family. If your travel plans must be changed, immediately notify your department and anyone who might be expecting you.
- Tag all luggage. Use your name and business address. Keep your luggage locked and never leave it unattended. There are luggage tags available that cover the face of the tag, so you have to lift up the face of the tag to read who it belongs to. This eliminates anyone from finding out who you are by casually glancing at the tag.
- Avoid using public transportation, such as the subway or bus. If you must use this type of transportation, select a busy, well-lit stop. Travel with a companion, if possible.
- If traveling by car, don't leave luggage or other valuables visible when the car is parked.
- Do not park your car on the street overnight; use a parking garage or lot. Lock your car and remove the keys. Leave nothing in the car.
- If everything fails and you find yourself in a bad situation, remain calm and attempt to withdraw from the situation without being noticed.

No matter where you are, keep in contact with your corporate security office and report any unusual incidents.

Follow these procedures:

- Contact the United States embassy or consulate (if no official U.S. representative is available, representatives from Australia, Canada, the United Kingdom, or another friendly country will usually help you). Contact a consulate if you are a crime victim, if you have been arrested by the authorities, or if you are injured or ill and hospitalized.
- Contact local authorities if that is deemed necessary by the official consulate representative.

- Contact your local office or designated legal counsel.
- Finally, contact the local corporate security office.

SCHEDULED AIRLINE SECURITY

Our country is trying to do something about the horrendous problem of terrorism, especially in the area of attacks on scheduled airlines. By no stretch of the imagination is this an easy task. There has been a recent flurry of legislative actions intended to combat both domestic and international terrorism, and many of these actions have been aimed at aviation security. The basic philosophy behind these moves starts with the U.S. Department of State's policy on terrorism, which, in the department's words, is:

1. We condemn all terrorist actions as criminal and intolerable, whatever their motivation.
2. We will take all lawful measures to prevent such acts and to bring to justice those who commit them.
3. We make no concessions to terrorist blackmail because to do so would merely invite further demands.
4. We look to the host government, when Americans are abducted overseas, to exercise its responsibility under international law to protect all persons within its territories, including the safe release of hostages.
5. We maintain close and continuous contact with host government(s) during an incident, offering support with all practicable intelligence and technical services, but we offer no advice on how to respond to specific terrorist demands.
6. We understand the extreme difficulty of the decisions governments often are called upon to make, for example, as a practical matter, how to reconcile the objectives of saving the lives of the hostages and making sure that the terrorists gain no benefit from their lawless actions.
7. International cooperation to combat terrorism remains essential, since all governments, regardless of structure or philosophy, are vulnerable, and we intend to pursue all avenues to strengthen such cooperation.

A prime example of U.S. anti-terrorism legislation is Senate Bill 2236 introduced by Senator Abraham Ribicoff. The bill seeks to upgrade security standards and provide a schedule for long overdue international and domestic security actions. In short, this bill provides for the formation of a U.S. Council to Combat Terrorism, with authority to impose sanctions on countries that aid and abet individuals or groups which commit acts of terrorism.

Once a country is earmarked as an offending state, the following coercive penalties would be imposed against it:

a. The president would be required to declare the country "dangerous for Americans to travel to or live in."

b. All commercial air service between the country on the list and the United States would be suspended. This ban would prohibit direct flights by the country's own carriers, third-party carriers, and U.S. carriers. It would also bar indirect flights between the country and the U.S. by both the country's own carriers and U.S. carriers.

c. No passenger who has traveled to or through a country on the list, or whose journey originated in such a country, would be allowed to enter the United States unless his passport contains a visa issued 1) by a third country in the third country, 2) after the date of his departure from a country on the list.

d. No plane would be allowed to unload in the United States if it is carrying any baggage including checked and transit baggage associated with any passenger referred to in (c) above unless the baggage has been thoroughly searched in the course of the procedures required for such passengers in (c) above.

e. No plane would be given permission to land in the United States if it has landed in any country on the list until it has been thoroughly serviced and inspected in a third country.

f. All persons, and all freight or mail, that have come from or passed through a country on the list within a year of the time at which entry into the United States is sought, would be subjected to thorough inspection before being allowed to enter the U.S.

g. No export licenses would be granted for the sale or transfer of items contained on the munitions list to any government-to-government and commercial transactions.

h. The sale or transfer of any nuclear facilities, material, or technology to any country on the list would be prohibited.

The president would be required to determine which foreign airports are less safe or secure than comparable facilities subject to U.S. safety and security standards. In making his determination, the president would be required to take into account all available data and any information obtained through relevant or necesary safety and security inspections.

The spirit and intent of this anti-terrorism act ties in with the business mission and security objectives of corporate/business/executive and charter flight operations. Denying a safe haven for the aviation criminal and terrorist is a basic deterrent to further violent acts for criminal or political purposes. These criminals will not respect either side of our borders. They will only respect our collective resolve to crush them.

What can you do as an airline user to avoid the problem of airline hijackings? If you are traveling from the U.S. and going overseas, choose a U.S. carrier. If you can fly direct to your location, do so. Try not to stop. Avoid layovers.

If you must change planes, do so in safe areas. The United Kingdom, West Germany and Japan are all safe places to stop. Wear inconspicuous clothing. Don't stand out in the crowd or wear expensive jewelry. Don't carry expensive pieces of carry-on luggage with you. Never carry anything that indicates that you are an executive with an American firm. Spend as little time as possible waiting in airports.

The best place to stop terrorism aloft is at the airline terminal. International airport terminals are the crossroads of the world. This is why they are used by terrorists in their misguided attempts to settle the arguments of the world. It is possible to produce a terror-proof airport, but it may require procedures and restrictions that travelers may not like. Passengers would have to submit not only to metal detectors but body searches. Your carry-on baggage would be limited to one piece and would be opened and meticulously searched, with an armed guard standing by. Checked baggage would also be opened and inspected. All of it. This would be required to stop terrorists from checking in luggage containing a bomb and then not boarding the airplane. All cargo and food entering the airplane would go through the same process. These security measures would mean a huge increase in labor costs (and therefore a major jump in ticket prices) and cause enormous time delays. Unfortunately, if terrorism continues to be a problem (and it shows no sign of letting up), this may well be the future of airline travel.

THE FAA

In the United States, civil air safety is under the jurisdiction of the Federal Aviation Administration (FAA). The FAA sets security procedures for airports, including screening and search procedures.

Each airport in the U.S. is equipped with metal detection systems. When originally installed in the early 1970s, many objects, including firearms, set off the machine. The latest state of the art in pulse-field technology can be set at a threshold that does not alarm at the passage of watches, belt buckles, and small metal objects. It can be set to detect guns, large knives, and other weapons. It can detect both ferrous (iron or steel alloys) as well as non ferrous material.

Active research is being conducted in the field of the detection of explosive devices, dynamite, TNT, and plastics.

The detection devices are augmented by personnel screening the carry-on baggage through X-ray machines. There are more than 700 X-ray machines in operation at U.S. airports. There is an armed uniformed officer immediately available, if needed.

When flying on commercial airliners, use some basic caution. Preboard if possible, or board just before flight time. This eliminates the need to mingle with the general public. Once on board, use common sense in your relationship with other passengers. Be careful not to disclose your identity, corporate connection, and travel plans.

INTRODUCTION

Personal security is not something you can do by yourself. Your family plays a big role in your security plans, and you must consider how your plans may affect your family. When living in a terrorist-infested environment, no other issue requires more cooperation from the entire family than family security. Different family members have different needs, so each member of the family must be considered separately when designing a personal security program.

As a typical parent, you'll be more concerned about the protection of your family than you will be about your own security. As a representative of American business abroad, however, you are at greatest risk, so your personal security plan will focus primarily on the risks you face. Terrorists rarely strike at families, but it has happened. You must do everything possible to insure that your security plan is a balanced plan, based on a realistic evaluation of the threats all of you face in a strange and potentially dangerous country.

PERSONAL SECURITY AND THE FAMILY

If you are living in a terrorist environment, your personal security requires the cooperation of everyone in the family. It is virtually impossible for you to be constantly on the alert for possible trouble. Your family can become several sets of eyes and ears for you. A word of caution: Be very careful about their role in security. They should never do anything that puts them in danger, but should always be on the alert for potential hazards. Living in a terrorist environment, your family will have to learn to help themselves as well as help you. Your personal security program will affect your family as well as you, and you cannot leave your family out of the decision-making process. Once the program is developed, have a family meeting and go over the details. You don't want to scare everybody, but you can never lose sight of the fact that they are a big part of the program, especially if something goes wrong. You must give them all the information that will be needed to help you and the company in the event of an emergency. Don't keep unpleasant truths from them in the interest of "not frightening them." A certain amount of fear can be a powerful motivator.

Your family's job will really be to support you, because in most parts of the world the danger to them is minimal. Until recently women and children generally faced a much lower risk from terrorism than men. This was especially true in most incidents of terrorism against business executives, and is still true in most locations around the world. Most terrorists, urban guerrillas, and revolutionary groups tend to avoid violent acts toward women and children. Killing women and children makes for bad press and bad public relations. Most terrorists adopt this "hands-off" policy toward women and children—most, but not all.

Acts of violence against women and children in most "civilized" societies are accepted only in a wartime situation. Unfortunately you *are* at war, at war with certain types of terrorist groups who see nothing wrong in gunning down or blowing up innocent women and children. Dealing with these types of groups requires extreme caution. If you are in an environment that has a recent history of random violence directed at the general population or, more specifically, against the general American population, then you must take every precaution possible to protect your family. The opposite is also true. If the terrorist group in your area is the type that will not attack women and children, it is good for your security program to have them around. While it may sound terribly cold-blooded, it is a fact that some terrorist groups will call off an attack on a man in his car or home if they feel his wife or children might be injured in the process.

If you have been selected as a target and have done good work on your personal security program, once you leave the house your spouse's threat level is very low. Since it's you they want, terrorists will not do anything to your family once you leave the house.

Trouble can come through your family's interactions with their friends, both American and local. Like you, your family should keep a low profile, and never discuss your personal security with anyone outside of the family or your company's security department. Sometimes your spouse may be proud of the status you have achieved with the company and want to brag about the fact that you are so important to the company that you need guards and around-the-clock security. This is not only foolish, it's very counterproductive and can result in disaster.

To the male reader: You must understand that your wife can play a large part in maintaining your low profile. No matter what she does, where she goes, and who her friends are, she must always play the same role as you; that is, she must at all times maintain a low profile. It will be much safer for everyone if she does.*

Working overseas creates new friendships. That is the good part of living overseas. We don't mean to make living overseas sound like living in a prison camp or high-security facility. Safely done, with a heads-up, eyes-open attitude, it can be a fabulous experience. Friends are very important when living overseas, but be careful who you and your family select as friends. Choosing the wrong ones could lead directly to disaster. Screen your friends and the friends of the family carefully. This can be a traumatic experience if you have children, who are so trusting. To help alleviate this, your family should make friends with Americans, at least at first. You should instruct all of the family to be leery of anyone who wants to be an instant friend, especially someone who wants to force a friendship on you. It's hard to strike a balance here: When does a person's friendliness become too friendly? It's a tough one to call, but most people's instincts are accurate. Paraphrasing an old aviators' adage: "If it *feels* bad, it *is* bad."

Explain to your children old enough to understand that they should tell you about their new friends. Be careful to find out if your children's new friends know more about you and the rest of the family than you feel they really should. The Marighella handbook trains terrorists to get as much information about you as possible. This can be accomplished most easily through people you think to be your friends.

Try to maintain safety in everything that you do. Once you have lived in the country, a simple task like shopping will become routine. Shopping should be done only in the major business and commercial areas. Stay away from the black market, back-alley areas of town. If there is a tremendous difference in prices (and there usually is), you and your family should still not do any shopping there. Let someone from your household staff do the buying for you.

*Patrick Collins, *Living in Troubled Lands* (London, England: Faber and Faber Limited, 1978), pp. 131.

Your night life also needs some thought. Go to restaurants and night clubs patronized by nationals, not to places solely frequented by Americans or other foreigners. Terrorists frequently target establishments that cater to an American crowd. Witness the April 1986 bombing of the La Belle disco in West Germany that killed and injured U.S. servicemen and precipitated the U.S. air strikes against Libya. La Belle was bombed because it was well known to be a favorite watering hole for local G.I.s.

Staying away from clubs like these must be an iron-clad rule in countries that have experienced violence against Americans in public places. As a rule, terrorists do not want to hurt their own people and will not bomb or machine-gun a restaurant loaded with their own people. Most of these attacks have been against places frequented by Americans. Know them and stay away from them.

Anytime you go out at night and you are not taking your car, consider taking a taxi. The taxi can offer you door-to-door service. Use a taxi company that you know is reliable and safe.

As mentioned in a previous chapter, you and your family must have a good relationship with the servants, domestics, and guards. If your servants don't like your family, they won't do much to help with security. If they dislike your family enough, they may even seek revenge by working with terrorists against you. Loyal help can be a tremendous asset. We cannot stress that enough.

PROTECTION OF CHILDREN

No matter where you live, overseas or in the U.S., you and your spouse have responsibility for the security of your children. Keep in mind at all times that they are children and that you must be responsible for all the security decisions regarding their activities. Moreover, you are not always going to be there to make decisions for them. It is your duty to instruct them on how to cope with situations in which they will have to make their own decisions.

Don't keep your kids uninformed; have regular, open discussions on personal security with them. Make sure they understand and respect your need for a personal security program. Although it may be difficult, you must review with your children their overall activities, and examine them for possible security problems. In the U.S., there are the usual security problems that all parents worry about—strangers and weird men in raincoats lurking near the playground, that sort of thing. In a terrorist environment, the number of potential threats increases dramatically. You may have to limit some of your children's activities. Treat them like adults at all times. Explain to them why they have to limit their activities. This is essential. Your children must know *why* these restrictions are being put on them. If they misunderstand, if they think you are being unnecessarily paranoid, if they think you are imposing these restrictions on them arbitrarily, or as a punishment, if they don't *understand* the threat, disaster can very likely result. This is something everyone in the family must understand. Life will be different living in a terrorist environment and will require sacrifices on everyone's part, including the kids'. You and the family must be willing to accept the difference.

Since the children will spend most of their time at school, find a school that provides not only good education but also good security. Unlike schools in the U.S., many schools overseas are very sensitive to security problems, and their administrators and teachers will be very receptive to your security suggestions. School authorities should be advised to contact a parent, or someone you have designated, prior to releasing the students to anyone during the normal school hours. They should always confirm the identity of the caller requesting an early release of the child. Inform the school that the child is required to return the call to verify that the caller is a parent or an authorized person. If there is any doubt as to the authenticity of the call or the caller, the child should not be released. If there is a school event that will take the children away from the school, they should always be watched by an adult. School personnel should be on the alert for suspicious people loitering in or near the schools. If their presence cannot be explained logically, the police should be notified and given descriptions of the person(s).

Advise your children to take the safest possible route home from school. The safest route is the one with streets most heavily traveled by other children, and which have controlled intersections with adult crossing guards.

When your children are not at home or school, they will be socializing with their friends. Be careful who they associate with. That bears repeating. *Be careful who they associate with.* Their activities outside the home should receive the same security considerations your security outside of the home receives. They have to be cautious of where they go and when they go there. The primary consideration is that of random terrorist bombings directed at the American community. Children can be victims of being in the wrong place at the wrong time. If there has been a series of bombings in public areas, curtail their extracurricular activity. Examine their outside activities carefully and decide which ones bring them close to danger.

No matter how old children get, parents tend to view them as youngsters. Most children are much smarter than most parents give them credit for. Most kids have the ability to make the right decision at the right time. After all, growing up isn't easy, and kids develop some pretty impressive survival skills along the way. The problem is that they don't usually know how to react to situations happening to them for the first time. No one does. In order for them to make the right decisions you must give them the proper information along with your confidence in their ability to make the right choice. If the family's personal security program is worked out properly, your children will quite likely enjoy the stay in a foreign country more than you will.

Encourage your children to talk with you about any problems or questions they have about security.

Alert your children never to approach or enter a strange car. You should be sure that the children understand that you will *never* send a stranger to pick them up from school or from a friend's house. They should always be picked up by people they know.

Encourage children to walk and play with other children and to avoid playing in vacant or deserted areas.

It is inevitable that when the children get older you will have to leave the home; even it is for a brief period of time, there are again some simple guidelines:

- When you are not home and the phone rings, instruct the children that they should not give out any information over the phone, especially their name or address. Under no condition should they ever tell the caller that they are home alone. The best advice for very young children is to instruct them not to answer the phone except when you call at a prearranged time.
- Around the age of four or five, children should know how to call for police and fire assistance. Be sure these numbers are on or near the phone in a clearly visible spot. *Never* leave the children without a person to contact. Leave a number where they can reach either you or a trusted adult. Be sure your child knows his/her name, address, and telephone number. Make sure they know who they may let into the house. No one else should be permitted to enter. Instruct your children to come straight home from school. They should not take shortcuts through alleys or deserted areas.
- It is an unpleasant thought, but you should prepare a plan of action in the event your child gets lost. Instruct your child's school to call you immediately if your child is absent.

SECURITY AND EXTRAMARITAL AFFAIRS

This is a tough subject to talk about but no personal security program is complete without bringing up the subject. Extramarital affairs are not common, but living overseas tends to create situations that can foster out-of-the-ordinary activities. Why these occurrences take place is not a matter of personal security. *How* they take place is.

We've already reviewed a case of an important man being killed on his way to visit his lover in the assassination of President Rafael Trujillo. There have been cases where executives were kidnapped while making the same journey. The most elaborate personal security programs can be reduced to ineffective garbage by the omission of a delicate little fact such as a regular trip you make that you would rather keep a secret. Such secrets can kill. Having an affair is your business. But don't compromise security for your extracurricular activities. If you do not adhere to a particular part of your security program because it will interfere with your love life, you are putting the safety of yourself and others that may depend on you in terrible jeopardy. Keep in mind that you have no secrets from the watchful eyes of the terrorist.

INFORMATION FOR THE FAMILY

Hope that it may never be needed, but you should put together a fact sheet on the family. When assembling your Family Fact Sheet, include the following for yourself and each member of your family, as applicable:

Full name, including middle name and nicknames.

Scars or special identifying marks, including those marks usually covered by clothing.

Date and place of birth.

Special medication requirements.

Details concerning second or vacation homes.

Locations and telephone numbers of employment, and/or volunteer activity, if regularly scheduled.

Names, addresses, and telephone numbers of all schools, plus names of school principals and children's teachers.

Names, locations, and phone numbers, including home telephone numbers, of instructors for recreational classes such as Little League, needlepoint, tennis, dance, and/or music classes, etc.

Regularly scheduled events such as hairdresser appointments, association dinners, etc.—including times, dates, locations, and phone numbers.

Auto registration, vehicle descriptions, and license numbers.

Names, addresses, and telephone numbers of important business contacts, banking relationships, attorney, doctors, dentists, pharmacists, and close friends.

List individual habits, language facility, extent of education, military service, hearing ability, degree of familiarity with firearms, athletic prowess, visual acuity (with and without glasses), degree of alcohol use, partiality to a particular color, personal nature, i.e., argumentative, casual, determined, humble, submissive, aggressive, friendly, trusting, skeptical, cautious; and other distinguishing characteristics, habits, or specific interests.

Include simple floor plans of residence and vacation homes, with addresses and telephone numbers.

Attach recent photographs of all persons listed. Include photos of eyeglass wearers, both with and without glasses, plus extra photographs of people wearing wigs if this substantially alters their appearance.

Information about senior family members should list grandchildren, including those in other cities.

Update the above information annually. Make copies and have them available at home (hide them in a secure place!), to nonresident family members, and to your office. Let your secretary have a copy and give one to your security director.

Security is something that all families need to discuss. Remember that the kids see terrorism on the TV nearly every night. To think they don't know what is happening is foolish. Discuss it with your kids. Let them know what you think about the problems and ask for their thoughts. Let them know what you are doing about the dangers of terrorism, and that you need their cooperation to fight it. Children are not immune from the anxiety caused by terrorism. An informal discussion between parents and children of terrorism's whys and wherefores can break down the stereotyped and often widely inaccurate images created by the media. Have discussions with the children explaining the problems and causes of terrorism. Finding the answers to the questions they ask will probably teach you a lot as well. Understanding the basics of terrorism is a family affair and something every family abroad should know.

FBI GUIDELINES FOR CHILDREN

Should a kidnapping attempt be made or an actual kidnapping occur, the security guidelines issued by the FBI in 1974 should be followed. These protective security measures are as follows:

A. *Safeguards for Children*

1. Keep the door to the children's room open so that any unusual noises may be heard.
2. Be certain that the child's room is not easily accessible from the outside.
3. Never leave young children at home alone or unattended and be certain that they are left in the care of a responsible, trustworthy person.

4. Instruct the children to keep doors and windows locked and never to admit strangers.
5. Teach the children, as early as possible, how to call the police; and instruct them to contact the police if strangers or prowlers are seen around the house or attempt to get in.
6. Keep the house well lit if it becomes necessary to leave the children at home.
7. If you have servants, instruct them not to let strangers in the house.
8. Schools, as well as parents and youth agencies, should take steps to make sure that adult supervision is provided in school and recreation areas.
9. Advice children to:
   a. Travel in groups or pairs;
   b. Walk along heavily traveled streets and avoid isolated areas, where possible;
   c. Refuse automobile rides from strangers and refuse to accompany strangers anywhere on foot;
   d. Use city-approved play areas where recreational activities are supervised by responsible adults and where police protection is readily available;
   e. Immediately report anyone who molests or annoys them to the nearest person of authority;
   f. Never leave home without telling their parents where they will be and who will accompany them.

All parents live in fear that something will happen to their children. During the last few years, kidnappers have unfortunately resorted with increased frequency to kidnapping the children of key executives of organizations they hope to blackmail. This terror tactic has proven very effective and is likely to become even more popular. Teach the following measures to your kids to help reduce their vulnerability to a kidnapper's threat:

1. Never let strangers into the house.
2. Avoid strangers and never accept rides from anyone that they do not know.
3. Refuse gifts from strangers.
4. Never leave home without telling you where and with whom they are going.
5. Learn how to call the police. They should be told to do this if they ever see a stranger around the house while you are away.
6. Travel on main thoroughfares, where possible.

7. Tell you if they notice a stranger hanging around your neighborhood or prowling around your home.
8. Play in established community playgrounds rather than in isolated areas.
9. Give a false name if ever asked theirs by a stranger.

Family members should be taught that strangers are a potential danger. Kidnappers will often try to use some pretext or other to enter their victim's home. For that reason, family members should be warned never to open the door to a stranger. Visits by maintenance personnel and the like should be scheduled. If a person claiming to be from a utility company or other such organization wants to enter, ask for ID, then make a call to the pertinent organization *before opening the door* to let them in. This is especially important in the case of unscheduled calls, but follow this procedure for both scheduled and unscheduled visits.

The distressed or injured motorist ploy is another one that has been used successfully in a number of kidnappings. It is the essence of simplicity. A stranger knocks on the door and says either that his car has broken down or that he has had a wreck. Often a woman is used to make the act more convincing. The "victim" then asks to use your phone to call for help. Once you open your door, you are at his mercy. Teach your family that if such a story is told to them, the proper thing to do is to keep the door locked and tell the stranger you'll call the authorities for them. Such a reaction may seem heartless, but it's the most realistic and best possible thing to do in such a situation. Such is today's troubled world.

SPECIAL PRECAUTIONS FOR WOMEN

Men carry wallets. Women carry purses. Men get their pockets picked, quietly, calmly, so gently they almost never realize it's happened until it's too late.

There's no such way to get a woman's purse away from her. That's why the crime is called "purse snatching." The theory behind the crime is simple: Grab what looks to be a purse containing some things worth stealing, and run like hell. As a woman, what can you do? Know a few basic facts.

Purse snatchers will go out of their way to hurt you if you resist. The general advice given to women walking down the street is to hold their pocketbooks close to their body and with both hands. Unfortunately, while this tactic may deter some purse snatchers, it won't stop most of them. The thief is going to catch you off guard, maybe even throw you down on the ground. Many of the women injured in purse snatches are hurt as they're thrown to the ground during the attack. Don't keep large sums of money in your pocketbook.

The following suggestions pertain primarily to actions taken to prevent criminal activity; however, these actions will cumulatively make it more difficult for a terrorist to attack. When followed, they increase the sense of security.

When shopping: Never leave your purse unattended in a shopping cart or on a counter. When asked for identification, give only the information requested. Never surrender your entire wallet or card case. Don't flash large amounts of money. Check credit cards periodically; immediately report any loss. Maintain a record of all account numbers and addresses of the companies to permit a quick reporting of lost cards, thus limiting liability. After making large purchases, check to see if you are being followed.

When driving: Look inside before getting into your car—an assailant may be waiting for you. Keep your car's gas tank full. Keep the windows rolled up and the doors locked. Park in a well-lighted area near your destination. If followed, blow your horn repeatedly to attract attention and drive directly to a safe location. If the car breaks down, raise the hood and trunk and remain inside with doors locked and windows rolled up until police arrive. Ask anyone who offers assistance to call the police. Don't pick up hitchhikers. At night drive only on well-lighted streets, if at all possible, even if it means going out of your way.

When walking: Walk on well-lighted, heavily traveled streets. Avoid shortcuts through alleys. Stay in the middle of the sidewalk. Hold your purse under your arm, with the latch on the inside. Be prepared to run if followed. If threatened from a car, run in the opposite direction to seek help. Use a well-lighted bus stop, and remain there until the bus arrives. Observe fellow passengers. If you can, sit near the driver. If you think you are being followed, find a police officer or call the nearest police station from a pay phone and report the incident; attempt to identify the individual. If you are approached by a suspicious person, cross the street or change direction.

When in an office building: Use an elevator where possible. Don't risk an attack in a poorly lit stairwell. Stand next to the control panel in elevators. If threatened, punch alarm button. Never leave keys or valuables in coat pockets. Use discretion in revealing personal plans to others. If working, keep your purse locked in a desk or file cabinet. When working late, notify the building security officer before leaving the office so he will know when to expect you in the lobby. Report suspicious persons or actions in your building to the security officer or supervisor.

If attacked: Scream as loud as you can. Strike back fast and aim for vital spots:

GOUGE EYES WITH THUMBS, SCRATCH WITH FINGERNAILS; SCRATCH ACROSS THE FACE WITH A KEY OR FINGERNAIL FILE; BASH TEMPLE, NOSE, AND ADAM'S APPLE WITH PURSE OR BOOK; JAB KNEE INTO GROIN; STOMP DOWN ON INSTEP; KICK SHINS; GRAB FINGERS AND BEND BACK SHARPLY; AND POKE UMBRELLA, COMB, FIST, OR ELBOW INTO MIDRIFF.

Report the incident as soon as possible. At work, notify the security officer in your building, or the local police as appropriate. At home, report the incident to the security officer and the local police. Try to give a good description of an attacker that includes color of hair and eyes, build, scars or tattoos, height, weight, and complexion. It is important that you report all incidents; failure to report may result in an attack on another person.

11

The Use of Lethal and Non-Lethal Force

INTRODUCTION

At some point in your personal security program you will have to make a decision on how personal this personal protection is going to be. How far are you willing to go to protect yourself? Could you go to the extreme and fight back with a lethal weapon if the need arose?

If you are living in an area of high terrorist activity, you will more than likely be surrounded by people who carry weapons. In this atmosphere you will need to understand, if not approve of, deadly weapons. In the U.S., the carrying of weapons is something that few businesspeople think of as part of their daily routine. You may be a gun enthusiast or the idea of owning a gun, much less using one, may disgust you. You have every right in the world to not want guns to be part of your security program, but you must accept the fact that in some countries, guns are a way of life and there is no way to avoid having them around you.

There are two schools of thought in personal protection and firearms. The first is that the police, your bodyguards, or the security guards around your home provide all the protection, and are therefore the ones that ought to carry the weapons. The second school of thought is that you are your own best protector, and a personal firearm provides the means necessary to supply yourself with the ultimate protection. Gun ownership is a serious affair. Before you buy a gun, you must be brutally honest with yourself. *If you abuse alcohol or drugs, if you have a violent temper, or are subject to long periods of depression, guns are not for you.* If you can honestly say "I could not take another person's life under any conditions," don't get a gun. Under either of these conditions, carrying a gun can be disastrous.

Whether you like guns or not, whether or not you choose to carry one yourself, if you are working in a terrorist environment you will need to have some knowledge about weapons, both lethal and non-lethal.

LETHAL WEAPONS

The most popular lethal weapon by far is the handgun. It is concealable and easy to carry. Before you decide to buy one, make sure that the laws of the country you are living in allow you to carry and own a gun.

Gun laws vary widely from country to country. You need to make sure that you are not in violation of the law by owning a given type of weapon. Be cautious, because the legal definition of a weapon varies from country to country. The easiest way to get the right answers is to contact the embassy in your overseas station. The embassy staff will be able to tell you the requirements of the law, how to go about getting a permit to carry a gun, and how to bring a gun into the country. Once you discover that owning a weapon is legal, you will probably have to purchase a gun in the country or buy the weapon in the U.S. If you want to buy the weapon in the U.S., you need to know how to get the weapon into the country. If you have a security department at your location, ask them to handle the purchasing of the weapon. They will know what to do and how to do it. Include them in the decision-making process. If they feel that you should not carry a gun or have one in your home, don't. They're the pros. Listen to their advice.

Buying a gun in the U.S. and sending it overseas can be a tremendous problem. If you send it by air freight, you may never see it again. If you try to put it in your luggage, you may have a hard time explaining to the customs people just exactly what you are doing with a weapon. The best solution to the problem is to let your company's security people take care of it.

Ammunition is another problem. Some countries will let you bring in a gun but, amazingly, no ammunition. There is a simple reason for this. Not many shippers will handle ammunition. Handling ammunition can be extremely dangerous. In this case, the best solution is to purchase a gun for which ammunition can be obtained locally.

The most important question you need to ask yourself is do you want to own a gun in the first place. If you have been around guns all your life, you will probably be uncomfortable if you don't have one while you're overseas, and will feel more at ease carrying a weapon. If you are not familiar with guns, carrying a weapon becomes a difficult decision. The decision to arm or not to arm should be based on your threat analysis. Even if your threat level is high, carrying a gun is still a very personal decision, based on some very personal questions. Will carrying a gun actually enhance your security? More than likely, the answer is no. A gun can serve as a deterrent. If terrorists know you are carrying a gun they may move on to another target, but they also may try to outgun you. The end result can be deadly.

If you are determined to carry a gun, make sure you know how to use it. Never own a gun without practicing with it. Make sure you know how to disassemble, reassemble and clean the weapon. If you are new to guns, make sure you can hit what you aim at.

Examine the average ambush scenario. The whole point of ambush is surprise, meaning the attackers have their weapons drawn and pointed at their target: you, in this case. More than likely you will not have time to draw your weapon. If you have the time to draw your weapon, you probably should have used that time looking for a path of escape. For kidnap prevention, a single handgun on one individual is an invitation to disaster. After all, your kidnappers are not there to kill you, they are there to abduct you. But if you come at them blazing away like John Wayne, their only choice will be to defend themselves. Given the amount and type of weapons used by terrorists, they can defend themselves, and very well, too. In a few brief seconds, both you and the terrorists will have failed in their mission. They failed in their kidnapping attempt, and you, lying dead on the ground, failed to defend yourself. *The best advice we can give you is that if you have never owned a gun, this is no time to start.*

Will you be able to use the gun should the need arise? It's impossible to answer this question before the fact. No one knows how they will react to a violent situation until it happens.

Once you've decided to own a gun, the next decision is what type to own. If the country does not allow you to carry a handgun, but you want something for home protection, a shotgun is a self-defense weapon worth looking at. Shotguns are legal in many countries. They are easy to use and generally safer than many other weapons. If you are going to get a shotgun, buy a five-shot version.

Shotguns are devastating weapons. If you are not familiar with the difference between a shotgun and a rifle, a shotgun fires a large number of small steel or lead pellets that emerge from the gun's barrel in a cloud or "pattern," as it's known. A rifle fires a single lead or copper bullet. As that bullet travels up the gun's barrel, spiral grooves or "rifling" etched into the barrel impart a spin to the bullet which stabilizes it in flight. This spin makes a well-made rifle amazingly accurate over long ranges of thousands of feet.

The shotgun's advantage is simplicity. That is also its major disadvantage. Since it fires a cloud of pellets and its barrel is not rifled (and it wouldn't make any difference if it was), the weapon is not accurate over long distance. At short range, however, it is extremely effective. Shotguns can be loaded with large pellets so that a hit from just a couple can bring down a target, many times without killing. At short range, however, shotguns kill and kill very well. While simple in operation, shotguns do require some training, albeit not much more training than any other weapon you would use.

There's a popular misconception about shotguns: People feel that they do not have to be aimed to be effective. Of course you must aim shotguns to hit the target. The tremendous advantage of a shotgun is its awesome one-shot stopping power. For home use, its major disadvantage is that the broad pattern of pellets spewing from its muzzle can easily strike unintended targets, such as members of your family. If you are going to keep a shotgun in the house, make sure you know how to use it and are familiar with the way it operates.

Most people consider the handgun as *the* primary weapon for self-defense. The two most important things that can be said about a handgun are: 1) Keep it out of reach of children; 2) Make sure you can get it if you need it. It's not of much use at the critical moment if it's locked in your safe. But leaving it in a location where a young child can get at it can be far worse.

Most people buying a gun for the first time buy a small snub-nose revolver just like the one Kojak uses on TV. If you buy a snub-nose, take time to become proficient with that gun. It is not an accurate or easy gun to use. If you're a beginner, you want a gun that is both accurate and easy to use. The conditions under which you may have to use the gun will require you to be very familiar with the weapon's operation. You may have to use it in the middle of the night when you've just been awakened from a sound sleep. Stumbling around in the dark is not the time to figure out how to use a weapon. So it has to be easy to operate. It should be heavy enough (meaning of sufficient caliber or gauge) that it can bring down the target with just one shot. A gun for home defense must be one of low penetration ratio to keep the slug from going through a wall and hurting a member of the family.

The physical size of a gun for home defense is unimportant. A gun for personal defense requires other considerations. If you are going to carry it on your person, it must be concealable. To be concealable, the gun does not need to be small, since a lot depends on where you want to conceal it. If you want to hide a gun in your briefcase, it can be pretty big. The news photos taken at the time of John Hinckley's attempted assassination of President Reagan showed a Secret Service agent toting an Israeli Uzi submachine gun, and a briefcase lying open on the sidewalk, revealing a cut-out foam interior shaped precisely to fit the Uzi. You probably don't need this kind of firepower, but the incident showed what a nondescript briefcase can conceal.

A gun for personal defense should be at least a nine-millimeter or .38-caliber weapon because the only reason to carry a gun is because someday you may have to use it. If that day comes you will need to get to it quickly and have enough power to stop the assailant with one shot. A nine-millimeter or .38 will do this.

What type of gun is the next decision. You have two choices: a revolver or an automatic. Both have advantages and disadvantages; it comes down to a matter of personal choice. The revolver is simple to operate and hardly ever has problems. Its disadvantages are that it is slow to reload and only holds six rounds. The automatic is fast to reload, easier to conceal, and can carry as many as fifteen rounds. But it is more complicated to operate and is more likely to jam due to dirt or lack of maintenance.

For the inexperienced person the best handgun is a .38-caliber revolver. This is the sidearm most police officers in the U.S. carry, so many that a well-known version of the .38 is known as a "police special." It is a simple gun to use, packs sufficient power, and works just about every time. If you already know how to use a gun, the 9-mm automatic with double action is probably the best selection. It offers speed, power, and a large ammunition clip. The key words here are "know how to use a gun." An automatic is not for a person new to weapons or a person who will not take care of the weapon. An automatic is the perfect weapon for the more experienced shooter. For the novice, a .38 loaded with high-performance hollow-point rounds is sufficient. If you have made the decision to use a weapon, make sure it is at least a .38 caliber and that you purchase a weapon of high quality with a reputable brand name.

Weapons may have their role in personal security but only if they are used properly. Here are some simple rules to live and stay alive by:

- Do not carry a gun unless you and your security department feel it's necessary.
- Do not keep a gun at home unless you can keep it there in a way that does not threaten you or your family.
- Do not point the gun at someone unless you intend to shoot them. If you don't have the commitment to fire the gun, pointing it at someone is the worst thing you can do. Your opponent could disarm you and use your weapon on you.
- If you have never used a gun and decide to carry one, you absolutely must gain proficiency in handling the weapon.
- *GUN SAFETY IS YOUR PRIMARY CONCERN.*

After you learn to take care of the gun, and know how to handle it safely, the next step is to learn how to fire it. In the U.S., there are many gun schools. Your local police department can help you find one. Training should be done with the gun, ammunition, and holster that you will be using all the time. The training will be useless if you use one type of gun, ammo, and holster in training and use another in the real world. If you are new to weapons, have your company find someone who can teach you how to operate the weapon.

NON-LETHAL WEAPONS

Non-lethal weapons offer an alternative to lethal weapons. Many of the non-lethal weapons available can offer some degree of protection. Their main attraction is that they can give your attack the element of surprise. In certain situations, you could use a non-lethal weapon to surprise your attacker and escape. They offer very little protection, and in some cases may actually aggravate the situation.

Chemical sprays like mace are a common form of non-lethal weapon. They work if sprayed directly into the assailant's face, but have not proven very effective against terrorists. There are two types of chemical agents, CS and CN, commonly encountered in commercially available tear gas guns. CN is a tear gas and CS is a choking gas. CS is much more powerful than CN. CN may have no effect on a person who is drunk, on drugs, or a person with a high pain tolerance, but CS will. The negative aspect of CS is that it takes about thirty seconds to work, while CN causes a burning sensation the instant it comes into contact with unprotected human flesh.

If you are trained with CS, it is by far the better option. Sprayed directly into an assailant's eyes, it makes that person change their whole attitude about the situation.

No matter what you use, you must be trained to use it. If your plan calls for protection against muggers or rapists, non-lethal weapons offer you some chance. But against an organized group of terrorists intent on kidnapping, the only thing it may do is get them upset. An upset terrorist can be an extremely unpleasant person.

PROTECTIVE CLOTHING

Wearing a bullet-resistant vest under a three-piece suit is not a common practice in the U.S. You may recall that President Ford began wearing one after the two assassination attempts against him in 1974 and 1975. The reason we know he was wearing one was that he did so before his suits could be retailored to accommodate the vest's added bulk, giving the president a somewhat chunky, overstuffed appearance.

In a terrorist environment you may find wearing a bullet-resistant vest as common as wearing a tie. To wear or not to wear is up to you. Once again, threat dictates use. If you suspect you have been selected as a potential target, wear a vest.

When buying a vest be sure to ask about *blunt trauma* and look at the results of any ballistics tests that have been conducted against the vest. The number of fabric layers and weaves are not only important to the amount of protection the vest provides, but also affect the overall comfort factor. How much the vest weighs is a major factor in deciding whether you will wear it or not. You may have the best intentions in the world, but if the vest is hot and uncomfortable, you probably won't wear it. Don't buy a vest unless you try it on, and any small adjustments are made before you buy it. Also ask the manufacturer if the vest is washable. Fabric armor is so susceptible to moisture that most manufacturers coat it with a waterproof material. Make sure you can wash your vest. If you can't, don't buy it. Carefully follow the directions when you do wash the vest.

There have been some great advances made in body armor in recent years. Most body armor is made of Kevlar® and is lightweight, cheap, relatively comfortable, and generally effective. Kevlar® is a trade name for a fabriclike material made by DuPont. According to DuPont, Kevlar® is five times as strong as steel on a pound-for-pound basis. When a high-speed bullet or fragment hits Kevlar, the fibers stretch, dissipating the impact energy. The U.S. Army believes in Kevlar® enough to require it as the major material for its new combat helmets.

The practicality of the Kevlar® vests for use by executives operating in a threatened environment is quite high. Wearing a vest is easy; you just put it on, secure the elastic straps, and then put on your shirt. It takes just a few hours to get used to the vest, after which it is easy to forget you are wearing one.

Under a shirt and jacket, the vest is almost completely indistinguishable. Without a jacket, it is possible to see a line here and there if you know what you are looking for, but it is certainly not sufficiently apparent for someone to easily spot. Thanks to comfort and concealability, a high degree of acceptance can be expected from those who may have to wear a vest.

It must be understood that these lightweight vests are designed to stop handguns, not rifles (there are concealable vests that can stop rifles, but these are quite heavy and relatively uncomfortable).

There is no reason for a threatened executive to use an inadequate vest when there are several on the market that can provide an extremely high degree of protection. No vest is completely bulletproof; there are exotic, armor-piercing handgun rounds capable of getting through virtually any lightweight vest, but these are quite rare and most people (terrorists included) are generally not aware of their existence. These special rounds are difficult to obtain in this country and almost impossible to find outside the U.S.

The most common threat to executives outside the U.S. is from handguns, submachine guns (which fire handgun bullets), and bombs. While the Kevlar® vests will stop shrapnel, they are not really useful against an explosive device detonated nearby. The most common handgun rounds outside the U.S. are the 9-mm and the .45 auto. In spite of the reverence with which many gun lovers regard the .45 auto, the penetration capabilities of this big, thick, heavy, but terribly slow round are so low that it is among the easiest to stop of all handgun rounds.

The 9-mm round is quite the opposite. It can be extremely difficult to stop depending on the bullet configuration, velocity, and construction—all of which can vary greatly. Because of the wide popularity and availability of the 9-mm—and in view of its high penetration capabilities, the ability of a vest to stop the 9-mm full metal jacket round should be considered as a standard for acceptance in the selection of a vest. This is especially important when the vest is to be worn outside the U.S., where this weapon is common and a favorite with the hijacking and kidnapping set.

A bullet will not knock you down. Contrary to what you see on TV and in the movies, and despite stories you may have heard, bullets don't knock you down. A simple rule of physics prove this: The energy received cannot be greater than the energy expended. This means that the force of a bullet striking a body cannot exceed the "kick" of the weapon when it is fired. What causes people to fall when shot is more their reaction to the bullet. If someone stuck you in the rear with a pin, you might jump into the air—it is not the "force" or "energy" of the pin which causes this, but your body's reaction to the sudden pain and/or its attempt to get away from it.

With the lightweight vests, a person wearing a vest that stopped a bullet will receive a bruise. There has been much discussion on the significance of "blunt trauma" injuries. Early critics of the Kevlar® vest claimed that while the vest might stop the bullet, the wearer would be killed by the bullet's blow—the "blunt trauma." Most of these critics have been silenced by the spectacular number of instances in which police officers' vests saved them from a bullet's penetration—without "blunt trauma" injury.

These new lightweight and concealable vests are indeed practical. They can provide the threatened executive with an unprecedented level of personal safety and security. And the vests are not simply effective against bullets; there is another dimension inherent in their use. Wearing the vest makes one *feel* protected. In many cases, this is most important.

As with cars, bullet-resistant vests can be purchased with varying levels of protection. Level I usually consists of 8 to 10 tightly woven Kevlar layers, offering protection against low-energy handguns from about .22 to .32 caliber; Level IIA typically comprises 16 to 20 layers of Kevlar, and withstands fire from medium-energy handguns such as a .38 special or a low-velocity .357 magnum. Level II, with 28 layers of Kevlar, protects against high-energy handguns such as high-velocity .357 magnums and most 9-mm machine pistols.

Levels III and IV join a Kevlar® weave with ceramic or steel plates to protect against rifle fire. Level III material withstands attacks from 223 Remingtons and other high-powered rifles, while Level IV protects against armor-piercing ammunition. These vests are very cumbersome to wear, however, and are mainly used for police work.

Notice in our discussion we do not use the popular term "bulletproof" vests. There ain't no such animal. These vests are not 100 percent effective, but in most situations they will protect a human torso against ballistic impact. A vest does nothing to protect your head, arms, or extremities. In a tragic incident not long ago a New York City police officer wearing a vest was killed in the line of duty when a bullet fired by an assailant entered his body through the vest's armhole and then pierced his chest. Most experts agree that use of the term "bulletproof" is unfair to vest wearers, giving them a false sense of security.

Body armor is now being made in the form of jackets, sport coats, and top coats. With these new advances, it is difficult for anyone to notice you are wearing it. Keep in mind that it's the comfort factor that makes or breaks the body armor. If it is comfortable to wear, and the threat level is high enough, there is no reason not to wear it.

If you don't want to wear an armored vest, there are a number of devices you can use to protect yourself. These devices are basically armored shields. The trade name for these devices is portable antiballistic devices. The most common of these devices is the armored clipboard. It is exactly what it sounds like. It is used by police officers all over the world, and can stop low-caliber rounds. The idea behind the clipboard is that if someone shoots at you, you put the clipboard in front of you to stop the rounds. Whether this device is effective or not depends entirely on the situation, on the type of round fired, the range, and, of course, how quick you can get whatever portion of your anatomy you're trying to protect behind the clipboard. The armored clipboard has certain uses, but we'd hate to be in a position where it was our only means of defense.

Then there's the armored briefcase, a standard briefcase with a piece of lightweight armor plate mounted on one of its inside walls. It too is designed to be used as a shield. When under attack, it can be held in front of you to provide some degree of protection. Another type of briefcase, one that opens into a complete shield, provides you with more protection, covering a greater portion of your body. This type of briefcase does not function as a briefcase, only as an armored shield.

The armored briefcase is worthwhile to consider. It can provide you with a degree of protection, is not expensive, and isn't much heavier than a standard briefcase. The bottom line on deciding among all ballistic devices from vests to briefcases is that you should listen to your security department. If they feel you need the device, get it and use it.

SELF-DEFENSE

There has been a tremendous increase in the interest in martial arts such as karate and related forms of self-defense. While there are many karate, jujitsu, judo, and aikido studios in the U.S., you will find very few in other countries. If you are interested in the martial arts as part of your personal security plan, find a training program that teaches you to defend yourself without becoming a kung fu expert. Remember that the martial arts are quite nearly a religion. If you want to get a black belt in some form of the martial arts, plan on taking a considerable amount of time. The martial arts can do a number of things for you. They are good for physical conditioning, mental discipline, and overall grace and coordination. If you are interested in them primarily as a form of self-defense, ask yourself this question: If the time comes, could you fight another person? For most of us, the answer depends largely on what the other person is about to do. If that person is about to do harm to your family, the answer is yes. Most people think of physical violence in the context of an argument, like when you were a child in the playground and the local bully picked a fight with you. That is not the situation you will face. A confrontation with terrorists will be much more violent than a playground brawl, the potential consequences much more grave than the odd bloody nose or two. If a criminal is in your home and about to attack you or a member of your family, you would be surprised at what you're capable of. Faced with the alternative of letting someone in their family get hurt, or killed, most people are quite capable of killing.

Self-defense programs can give a tremendous edge in what could easily be a life-or-death struggle. Your self-defense program should be custom designed for you. Understand that the success of any self-defense program depends on the instructor. Select one with experience dealing with executives, a high level of proficiency, and the ability to convey this knowledge to you.

HIGH TECH NON-LETHAL WEAPONS

Technology has created a new non-lethal weapon, the stun gun. If properly operated, the stun gun can put an assailant flat on his back in an instant. The stun gun is a viable alternative to small-caliber handguns, comprised of a self-contained rechargeable nickel-cadmium (NiCad) battery rated at 9 volts (but which produces a nominal 7.2 volts). Through microchip circuitry, the 7.2-volt potential is stepped up to between 40,000 and 50,000 volts. At the same time, the battery's 0.3-ampere current is reduced to a non-lethal 0.00006 ampere (0.06 milliamp). The high-voltage, low-current charge is stored in the stun gun's capacitor bank until, with a push of the thumb switch, the charge is pulsed at low frequency and is released swiftly through the twin probe electrodes into the attacker's body to disrupt neural and muscle functions.

In addition to its shock value, the stun gun can serve as an effective intimidator, even though it is not in contact with the adversary. The stun gun has a very short range; the target has to be close to you. It's a good device for warding off common criminals but certainly not a group of terrorists. As with all non-lethal weapons, it should be used with caution.

PROFESSIONAL BODYGUARDS

Thanks to the explosion of violent crime and terrorism throughout the world, one of the fastest-growing industries today is the protective services industry and along with that, bodyguarding.

When you think of a bodyguard the typical image is of an illiterate, muscle-bound man who has a hard time maintaining a conversation with a five-year-old. Nothing could be further from the truth. Just take a look at Secret Service agents, who make up the premier protection agency in the world, and you will find intelligent, well-trained individuals who simply exude professionalism.

If you decide you need a bodyguard, you are adopting one of the most advanced personal security systems by far. If the threat level is high and your security department recommends the use of bodyguards, use them. Overseas and in the Third World, bodyguards are very popular—not very well trained, but popular. Most chief executive officers should consider the use of bodyguards when traveling overseas, or at special functions such as stockholders' meetings. In a terrorist environment, bodyguards are usually a necessity, not a luxury. Although the actual hiring of bodyguards is a task for your security department, your new pals will be with you most of the time, so you should have a say in who gets hired. When hiring bodyguards, look at a number of issues. The first issue is training. Good training is important. There are a number of so-called "bodyguard schools" that claim to train an individual in one week to be a bodyguard. That is simply not realistic. These schools may be able to take an ex-police officer and give him some basic skills, but it is highly unlikely that they could turn him into a professional bodyguard. Look for former Secret Service or State Department personnel as protectors. They have the experience to go along with the knowledge. If you're doing the hiring, look for the following traits:

- Good moral character, with a higher-than-average degree of integrity. Check their past employment. They should receive excellent recommendations.
- No history of narcotic, drug, or alcohol abuse. The alcohol part is important. You don't want the person responsible for your protection to have a hangover.
- No criminal record other than minor traffic violations.
- Physically fit, and in generally good health.
- Good hearing and eyesight.
- Disciplined, patient, and thorough.
- Can work well with others.

Try not to guess if the protective bodyguard has these capabilities. Test them. There are very reliable physiological testing programs that you can put your applicants through that will give you an idea of how they match up to the above criteria. These people will bear the responsibility of protecting you and your family. In no case should you randomly pick a person just because he looks tough. Moreover, physical condition and appearance are far from the only considerations. Since they will be spending a lot of time with you and your family, it will be much easier for all involved if the personalities of your bodyguards complement those of you and your family.

Every person involved in protective services should "know the enemy"—a good background in the overall terrorist threat and an understanding of violent criminal activities are necessary. This includes knowledge of terrorist tactics, and how they may vary from area to area.

The candidate should be interested in physical training, and should be in good shape. He should exercise regularly. So should you, for that matter. Companionship in working out is an important fringe benefit of having a bodyguard. Now you've got someone to go jogging with! Since the bodyguard has to stay near you anyway, he might as well suit up and work out right along with you. President Jimmy Carter used to delight in trying to outrun his bodyguards during his regular jogging sessions (he never did).

Bodyguards should have experience in unarmed combat. Although having a black belt in judo or karate is desirable, it is not essential—the bodyguard just must be able to fight effectively and well. The person must know what to do in a street encounter or attack situation. People that have been trained in the classic martial methods may be very good at fighting, but a protector also needs to know disarming techniques, how to take a weapon away from someone quickly and safely.

A bodyguard requires excellent proficiency with handguns, shotguns, rifles, and, occasionally, submachine guns. This training is crucial. There is no way to "karate" either himself or his client out of a surprise terrorist attack or kidnapping attempt. Firearms are absolutely necessary. This does not mean that the bodyguard must be a competitive shooter, but he must achieve and maintain a functional combat capability with firearms.

As for handguns, he must be able to respond instantly to attacks at distances of seven to ten yards or less, and have excellent skills with aimed fire when time, lighting conditions, and cover permit such shots to be made without danger to the person being guarded. He must have comparable proficiency with the shotgun and rifle. Countersniping ability is often desirable. There are special combat shooting courses available. If the bodyguards have not taken one, they should strongly consider it.

Special driving skills are also a prerequisite. A bodyguard should not be used as a combination chauffeur/bodyguard. However, driving can be part of the protective team function. The driver must be trained in evasive driving, defensive and counterambush maneuver skills. A bodyguard should also be familiar with armored vehicles and cars, including their capabilities and handling peculiarities.

The bodyguard also has to be proficient at emergency first aid/CPR. This can all be learned absolutely free from the American Red Cross in the U.S. Check the Yellow Pages for the chapter in your city. Overseas, ask the local U.S. embassy where first aid is taught. Remember that a bodyguard's purpose is to save and protect lives. Besides the standard first aid, a bodyguard must know how to deal with gunshot wounds, stabs, etc. The Red Cross's facilities are excellent, as are its teachers. They certify you, and they have had more experience in teaching their subject than all the bodyguard schools in the world.

Someone in your protective program must know how to set up security in advance. If you are going to another country for a business trip, there should be someone in the group that can go in advance and set up all the security arrangements. This requires special knowledge that only comes from experience. This is also the advantage of hiring ex-Secret Service personnel. They are trained in this kind of advance work and are very good at it. Also someone in your protective program needs to have knowledge of such subjects as electronic security, alarm systems, and so on.

If you are employing more than one bodyguard, then you must designate one as team leader. This individual must have an above-average level of intelligence. He should be articulate and possess the ability to mix easily with any type of gathering. You will need this person with you in both business and social settings, and that person will need to easily fit into both environments. We've put down this gentleman in the past but in this case we really are looking for somewhat of a James Bond type. Not quite so suave as Bond, and definitely someone who can keep his mind off vodka and women better than Bond, but a bodyguard with that sort of quick mind, who keeps his wits about him when everyone else is losing theirs, is indeed a life saver.

Accordingly, mental agility of the subject must be well above average. A college degree is not a necessity but a college background is desirable as an indication of initiative, drive, and desire, especially if the academic pursuits occurred after military service or during employment.

It may surprise you, but youth is a distinct disadvantage for a bodyguard. The minimum age is 27-30 years old, with no defined upper age limit, since that limit should be dependent solely on ability and physical condition. The physical size of the agent should be approximately that of the protectee. A medium, athletic physique with coordination and speed is most desirable. The physical appearance of the agent is vital. The person must be able to blend in with your environment. You and the person you select should even be able to blend well with each other. Therefore, if you wear a three-piece suit, the agent should look and feel comfortable in the same.

The candidate's prior experience and background will determine skill levels. The person you have with you all the time should have prior police, military, and/or security backgrounds, but this type of background doesn't automatically make them suitable for the job. They still need training as a bodyguard.

The decision to employ a bodyguard is a complex one. Such a decision involves considerable economic investment, but, more importantly, it will have a dramatic impact on your lifestyle and your family. Adjusting to a bodyguard can be a traumatic experience for you and your family. Make sure you select a person everyone can get along with.

Pay is very important. In today's economy and to attract the caliber of person necessary to do the job, a starting salary of $35,000 a year along with allowances for clothing, lodging, and meal expenses is not unreasonable. This is for a first-rate bodyguard with good administrative skills.

This bodyguard selection process is extremely important. It can fail, however. Nothing is foolproof. You can have a bodyguard and still be kidnapped. The most important issue is to avoid bodyguards that are not qualified. Terrorists are not fooled by make-believe bodyguards. The only thing you'll accomplish having inadequate bodyguards is to create a situation that can get them killed.

The skills required of a bodyguard are many and are only attained through considerable economic and physical effort by the individual. Once acquired, those skills need constant honing to keep them effective. In today's volatile world, the task of providing protection has been elevated to an exacting science. If you are going to hire a bodyguard, make sure of his skills and character. If you select the wrong person, both of you will lose.

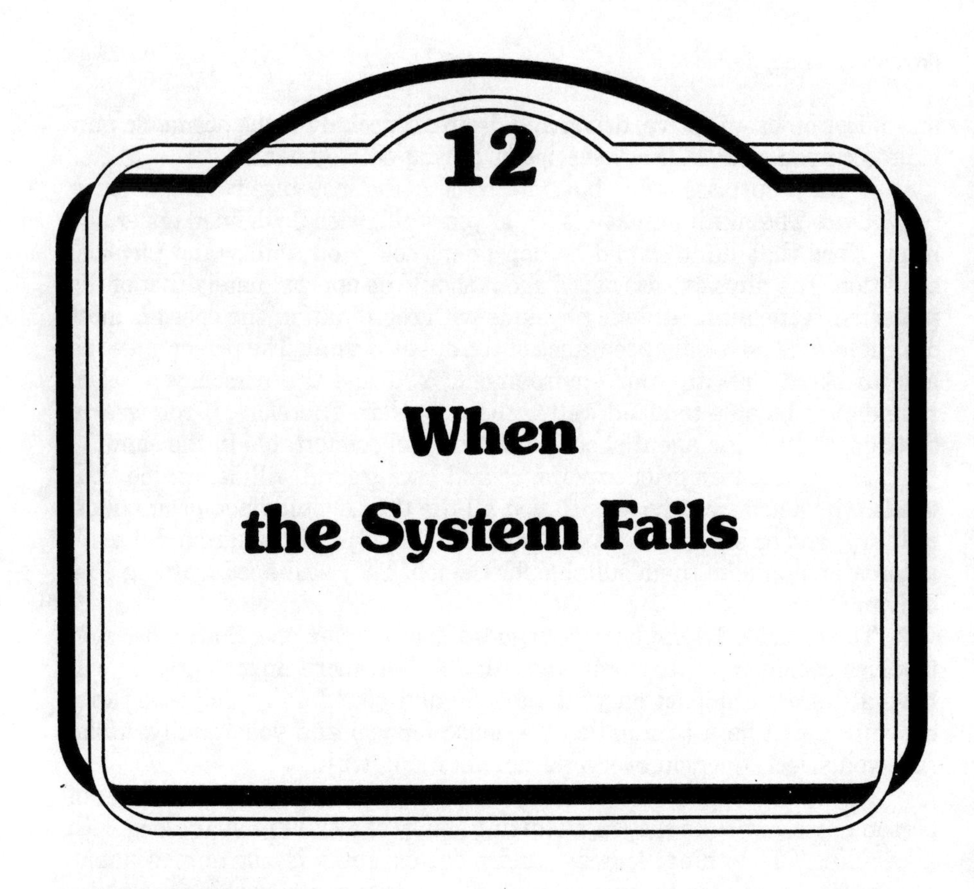

INTRODUCTION

When working in a terrorist environment you have to face reality, and the reality is that there is a very real possibility of you being kidnapped. Although not likely, it is not impossible. You must face the fact that it could happen and that it's in your and your family's best interest to make plans for the worst. Making these plans will not be a pleasant task. They are similar to the plans you would make in the event of a natural death. In a terrorist environment, it is of utmost importance that you gather the family together and discuss what needs to be done. There are some special issues that will need your special attention.

GETTING READY

First you need to create a will. You should have one anyway, no matter where you live. The will should be filed with a lawyer or relative in the United States. You must have a certified copy with you in your overseas location.

The most important document your family will need in the event of a kidnapping is power of attorney. This document will give your spouse the authority to handle your legal affairs while you are being held captive. Understand that in the event of you being kidnapped or killed, your family may not be in a condition to do much of anything, so it is imperative to also have your lawyer, or a close relative in the U.S., have power of attorney. A trustworthy representative in the U.S. may be better able to handle the legal problems than someone in the country you are working in.

All important papers and documents must be stored in a safe place. Make sure everyone in the family knows where they are. These papers should include any legal documents, trust, wills, powers of attorney, deeds, titles, insurance papers, stocks, bonds, bank records, and so on. Anything that you feel your family may need in an emergency should be included.

If you are kidnapped and held for a financial and not political ransom, money will become very important to the family. The money situation is far more important if you are kidnapped than if you are killed. If you are killed your family may be forced to live on your bank account until your will is probated. If you are kidnapped, they may have to survive on these funds for the entire time you are held in captivity. It is extremely important that your family have immediate legal access to these funds. Prepare for the worst and be certain there is enough money in the bank to support your family for several months.

The next issue is your health and personal records. It's important that you have a complete and accurate set of medical records with you at your location.: Your medical records provide the people who will be negotiating for your release with an indication of your current medical condition. If you need special medication, it should be noted in the medical records, as well as how long you can go without the medication. This can keep you alive if you suffer a prolonged captivity.

Once the basics are complete, you will need to do some plannning with your family and your company. Most companies have crisis management teams comprised of security, financial, and legal teams to handle problems like this. If your company does not have a crisis management team, put one together. No matter how efficient, effective, or helpful the company's crisis management team is, there are still some questions that only you can answer, questions such as: If you are kidnapped, should your family stay in the country or return to the United States? This needs to be discussed with your spouse and the family in advance. No one other than the family should make that decision. The decision can give you peace of mind. While in captivity you'll know where your family is and that they are being taken care of. Decide in advance if your family should try to raise the ransom on their own or go to your company. Raising the ransom demand is usually the company's job. More than likely, the terrrorists' demands will be made against the company. Also, you must consider whether or not you want your family involved in negotiating your release. In most kidnappings the company will take care of that for you, but your family should have some access and input into the negotiations.*

What do you want your employer to do in the event of your kidnapping? Since the demands will most likely be made against the company, it will have a good deal to say about the ransom payments. Find out what assistance it will give your family while you are being held. In kidnapping situations, most companies will give your family a great deal of support, and you may want the company to handle your financial affairs and to conduct negotiations.

Do you want your family to talk to the press? Decide in advance. The best advice is to let your company or the police decide if they should talk to the media. The decision in most cases is not to let them.

If your family is close to any of the country's political figures, should they contact them for help? You may have been kidnapped in the first place because you had such connections, in which case it may not be a good idea to use them.

If you are unfortunate enough to be held for a long period of time, what do you want your family to do? Should your family encourage the police or other authorities to release you by force? History suggests that hostage rescue attempts by local police generally result in disasters. Your family should never agree to a rescue attempt, unless it is certain that your life is in imminent danger and all other options have been exhausted. At some point during the kidnapping you may be required to communicate with the negotiators. When ransom is about to be paid, the payers will probably require proof that you are still alive. Knowing this, it's a good idea to develop code words you can use when contacting your family or the negotiators. Code words can be developed to let them know you are safe, the general location of where you're being held, and/or if the kidnappers intend to kill you whether the ransom is paid or not.

EVACUATING THE COUNTRY

Another unpleasant topic you must address before the fact is the possibility that you could conceivably be required to leave the country in a hurry. If this occasion should arise, you and your company must develop on evacuation plan. The plan should identify the evacuation phases and what is to be accomplished during each of them. While every company's needs are different, we offer the following evacuation outline as a guide:†

*Patrick Collins, *Living In Troubled Lands* (London, England: Faber and Faber Limited, 1978), p. 142.

†Brian Jenkins and Anthony Cooper, *Terrorism* (Stoneham, MA: Butterworth Publishers, 1984), pp. 164-66.

1. *Alert.* In the alert phase, personnel will be advised of the possibility of evacuation within the near future. They should begin advance preparation for a possible move and hold themseles in readiness to receive further instructions. This is a precautionary move from which they may either proceed to other phases of the evacuation procedure or from which they may stand down if the potential emergency is deemed to have passed.
2. *Readiness.* During this phase, personnel should be prepared to move the instant they get the word. Packing of possesions should be complete and all personal affairs should be in order. This phase encompasses all the work that must be done before the actual evacuation begins.
3. *Movement.* During this phase, personnel and their dependents will be moved from homes and offices to prearranged staging areas and then onto sea, air, or land exits. Staging areas should be selected with care and clear instructions given on how to reach them. Make sure your people understand those instructions. Arrangements for reception and care as well as transportation should be carefully spelled out well in advance.
4. *Transit.* This is an intermediate phase, involving much upheaval and a measure of discomfort. The primary concern is to remove personnel from the danger area as quickly as possible. Arrangements have to be made for transportation, care, and shelter, along with reception at designated destinations outside the area evacuated.
5. *Resettlement.* This will generally mean repatriation, although in some cases employees may simply be moved to other, safer areas where business may be continued. Arrangements for housing relocated personnel and their integration into the business organization elsewhere, as well as compensation and other adjustments, need to be considered during this phase.

LIAISON WITH THE AUTHORITIES

Unless there has been a complete breakdown of government (which is likely, otherwise why would you and your company be evacuating your people from the country?), an orderly evacuation can *best* proceed if certain necessary formalities are observed. Customs and other fiscal clearances will probably be required and movements may for many reasons be restricted or subject to official scheduling. An evacuation can only proceed smoothly if these matters are competently attended to by persons familiar with official routines and requirements.

In the event that all government breaks down, as happens in countries in which the government suddenly changes hands, decisions as basic as whether to stay or leave must be made, based on the best possible information. Situations of political upheaval have included the crises of governments in Viet Nam, Cambodia, Nicaragua, and, most recently, the Philippines. Each of these situations was different, and the reaction from American business on the scene was different in each. Know who and what you're dealing with. Make realistic risk assessments. Good luck.

PERSONNEL REMAINING IN POST

Circumstances may indicate that it is only necessary to evacuate some personnel, such as dependents and non-essential employees. In this case, the plan should include special provisions for the safety of those remaining behind and may include special living and travel arrangements, as well as increased security.

Time is crucial in evacuation operations. A major objective should be to secure as much time as possible for the accomplishment of each of the tasks assigned to the various operational phases. In planning, it's easy to create an evacuation plan that runs smoothly on paper. A paper plan always stays on schedule and works beautifully. In the real world, things screw up. Everything seems to take much more time than has been allowed for it in the planning process. Racing against the clock heightens anxieties and, in the worst of cases, induces panic. Allow for breakdowns in the situation while still in the planning phase. Be generous in your estimates of how long certain tasks will take. In the readiness phase, it's far better to be a little ahead of schedule than behind. It's also important that time, when available, not be wasted. Careful scheduling and close supervision of all phases of the evacuation are necessary to insure the correct degree of urgency. Too much time can be as great a bane as too little. The best amount of time is that which will allow the job to be done thoroughly without undue tension being forced upon those concerned.

Know that government does not necessarily work faster in a crisis. Allow the appropriate amount of time for obtaining clearances, exit permits, and the like. Timing, knowing when to make certain decisions and initiate certain actions, is no less important.

Above all, the human element in the evacuation of business personnel should receive constant sympathetic attention. There must be genuine sensitivity to what is involved for those affected by these events. These are people whose business and personal lives are being grossly disrupted by events over which they have no control. For many, disruption may strike with all the swiftness of a natural disaster. After the "heroic" phase, in which everybody tries to cope with the problems at hand, comes a period of adjustment and reflection. An evacuation is capable of generating strong emotions and deep resentments. It can result in loss of faith, lasting suspicion, even lawsuits. Often overlooked are the psychological damages suffered even by those who have made it safely out of the danger area. Counseling and other help may be needed and ought to be generously provided by any business truly interested in the welfare of its personnel. Much can be done to relieve the severe emotional stress induced by an emergency evacuation. Continued useful service by personnel affected may well depend on appropriate measures being taken in timely fashion.

PERSONAL RELATIONSHIPS DURING THE KIDNAPPING

During your period of captivity the negotiator will face the enormous challenge of trying to get you released, interacting with a number of people. While you're in captivity you will feel much better if you know what is going on back at the office. You need to know how the process works.

Your company will hire a negotiator to start working on your release. The negotiator will keep your family as calm and as unworried as possible, keeping in constant communication with them to inform them as to what both he and the company are doing and assuring them that the company will do everything possible for your safe release and speedy return. He may help your family by lending a hand with their day-to-day routine, trying to have them resume normal activities as much as possible. He will show them he has a great deal of confidence in his own ability to negotiate your release.

If possible, your family should not be interviewed by news media. Younger family members, especially teenagers, can innocently reveal detailed information which might jeopardize the success of the release. If necessary for their safety, comfort, and security, your family may be moved to a safe area, for instance a motel or safe house in a secluded location, until the situation is resolved. The police will usually cooperate by assigning protection to the family. If not, hire a private bodyguard(s).

As soon as practical, the negotiator will notify law enforcement of the situation. Most terrorist situations come under the joint jurisdiction of national and local authorities. Contact the U.S. embassy as soon as possible. The local law enforcement agencies will try to enlist as much help from them as possible. Your company will need to be careful about local laws regarding ransom payments.

The negotiating team will be and must be honest and frank with local law enforcement. In all situations, law enforcement will cooperate with your company and with each other for your safe return. Law enforcement agencies will do nothing to jeopardize your safe return.

Your company must honestly assess the capabilities of law enforcement. In some communities, local police have limited capabilities and experience. In such cases, it will be far better to notify national police to enlist their primary invesigative capabilities and to make sure that a competent investigation is conducted.

Negotiators must cooperate with other negotiators so that everyone concerned is aware of the situation. They must control all information released to the media. It will better your chances of a safe return if as little detailed information as possible is released to the press, television, and radio. It's possible to admit that a situation exists without revealing any details about the family situation, the amount of ransom demanded, or the details of the ransom delivery. Furnishing too many details will permit:

- Sick, antisocial, greedy people to enter the picture and complicate things with fraudulent attempts to get money (this is another reason why codes are necessary in all negotiations with the kidnappers);
- Overly aggressive news reporters to complicate funds delivery by maintaining close surveillance;
- The efforts of law enforcement to secure the return of the victim, to apprehend the kidnappers, or to recover the ransom funds may be jeopardized by releasing information that should be concealed. In a terrorist situation where the safety of personnel is at stake, it is better to release too little rather than too much information. Law enforcement advice should be sought about the release of specific details. Criminals and terrorists read the papers and listen to news broadcasts, probably more often and more closely than anyone else.

One person should handle all contacts with the media, and only that one person. All others involved in the negotiations should refer any contacts from the media to that one person. This will prevent the media from playing one official against another to obtain more information. It's a standard media procedure to attempt to persuade someone to give in to the temptation to feel important by releasing small details that are then used to confront a second person with the released information.

When negotiators are dealing with the terrorists, they will try to make sure you're alive. If it's possible for you to talk to the terrorists, tell them honestly:

- That your company will do everything for your release.
- That they just want to make sure that you are unharmed.
- That they would like you to write a note to them or to say something to them that only you would say so they know positively that you are alive and unharmed.

Ask one of the terrorists to refer to himself by an agreed-upon code name so that negotiators will know they are talking to the same person each time. It is not uncommon for several different people to try to collect a ransom in any publicized kidnapping. Each person calling should be given a different code name.

Obtain and repeat instructions for funds delivery. There may be attempts from both sides to haggle over the amount of your ransom. If the opportunity arises, they may tell the terrorists that they can obtain one-half or one-third the ransom amount in return for immediate same-day delivery to any spot the terrorist wants but that delivery of the full amount might take longer since local company officials must get authorization from higher-ups.

CREATING AN EMERGENCY BIOGRAPHICAL DATA SHEET

If you are kidnapped, it's very important that a complete record of your personal information be immediately available for law enforcement use. Your complete biographical dossier must be maintained in one central location for use only in the event of an emergency. Information should include color photographs of you and your family, your and their fingerprints, signature samples, and voice tapes. This information should be kept in individual envelopes sealed by the executive and updated by him on a regular basis.

The biographical data included in the file should, as a minimum, include:

- Complete name(s);
- Addresses of primary and secondary residence, as applicable;
- Personal telephone numbers;
- Complete physical descriptions: distinguishing physical features such as scars or other identifying and unique physical characteristics;
- Banks where money may be deposited, has been deposited, and where withdrawals may be made;
- Name of local physician, dentist, optician;
- Family cars: state license numbers, make, model, year, and color, and vehicle identification numbers;
- Schools attended by children;
- Names, addresses, and telephone numbers of immediate family or other relatives who could be contacted regarding whereabouts of the family;
- Credit card companies and card numbers;
- List of boats, campers, or other recreational vehicles;
- Profile of hobbies, clubs, and other activities in which family members participate.

CRISIS MANAGEMENT TEAM

If you are working for a major corporation they will more than likely have a crisis management team on line. While you are held in captivity it will help for you to know that there is a team of your peers working for your release. Members of the crisis management team will include individuals who have authority to implement and carry out the policy dictated by the board of directors and the procedures contained in the company's crisis management plan. The presence of more than five people on this team could easily lead to confusion at a time when confusion is least desirable. In addition, the team should be aware that all of the resources in terms of manpower and material of the company are available for their use on an *ad hoc* basis.

Members of the Crisis Management team should include:*

1. The Coordinator, Chairman of the Board, and Chief Executive Officer
2. President and Chief Operating Officer
3. Executive Vice-President of Finance
4. Executive Vice-President of Operations
5. Executive Vice-President of Administrative and Technical Services
6. Director of Security
7. Public Relations

Other members of the team may vary, depending on the nature of the threat or demand, or the location of the crisis.

In addition, the coordinator shall be responsible to:

1. Formulate plans and procedures for handling crisis situations;
2. Gather an advisory staff (if deemed appropriate) to generate information and perform services to facilitate these procedures. Example: A member of the legal staff may be necessary to review the plans for compliance with established corporate policy;
3. Maintain in a secure place the current Crisis Management Plan and Procedures;
4. Communicate these plans and procedures to authorized persons only and to follow up and insure that these individuals are fully cognizant of any changes in plan or procedures;
5. Maintain current personal information and biographies of all corporate executives in a secure place. The personnel department maintains a very limited amount of biographical information pertaining to company executives. Enclosed with this document is a biographical inventory. We recommend that each executive complete the document and that it be placed in his individual personnel file where it can be quickly located in the event of an emergency;
6. Staff and train the personnel necessary to carry out the crisis management program;

*James F. Broder, *Risk Analysis and the Security Survey* (Stoneham, MA: Butterworth Publishers, 1984), pp. 212-14.

7. Exercise good judgment in determining the course of action in a crisis situation not covered by approved policy;
8. Implement plans and procedures according to the following management plan.

The purpose of a crisis management center is to serve as the focal point for directing a coordinated and planned response during a crisis situation. It should be located within the organization's headquarters facility at or in the executive conference room. The center should be furnished with all documents, supplies, and communications that may be needed during a crisis. As an example, items such as tape recorders, office equipment, typewriters, and a log to record all calls and actions taken will be necessary, at minimum.

When an executive, employee, or family member becomes the victim of a kidnapping, or the company becomes the victim of an extortion or terrorist plot, the response of the organization will signal the implementation of the Crisis Management Plan (CMP). To implement the plan requires the authority to do so, and this authority must be clearly spelled out. Implementation criteria should be defined and some basic questions answered:

1. Who has the authority to implement the crisis management plan?
   a. Chairman of the Board and Chief Executive Officer
   b. Alternately, the President and Chief Operating Officer
2. The minimum circumstances that must be established for this authority to become effective (example: if a threat of kidnapping is received as opposed to news of an actual kidnapping)
   a. How long this authority will remain in effect
   b. Succession to this authority should the head of this authority be removed or incapacitated

When an extortion demand or threat is received it should be immediately reported to the decision-making authority as outlined in the CMP. The decision as to when the CMP should be implemented will depend on verification of the threat as true or false, and analysis of the threat posed to the company or its personnel.

13

Surviving the Hostage Incident

INTRODUCTION

Although the possibility of you being kidnapped and becoming a hostage is slim, as a resident of a foreign land and/or terrorist environment you must face the possibility. Your reactions during your kidnapping, and while being held hostage, will determine how your kidnapping will end. There is no doubt you will be in a state of shock during your kidnapping. You will become disoriented and easy to manipulate. To lessen the dangers inherent to a hostage situation, you must be prepared to capitalize on errors made by your kidnappers. Try to know in advance how you will act. Being prepared is the greatest asset you can have.

Understand what will happen. If kidnapped, your life changes dramatically in a mere matter of seconds. As a businessman you lead a busy life; you make things happen. When you are kidnapped, your life of activity will change to one of isolation and inactivity, discomfort and degradation. The kidnapping itself will be quick, probably violent, and certainly traumatic. There is no question that you will suffer some physical abuse. The attack itself can be catastrophic. Bodyguards or a driver may have just been killed. Whether or not they are killed depends on how the terrorists evaluate their ability to prevent the kidnapping. If terrorists feel that they can easily overcome any opposition your bodyguard(s) may present they will not kill them. If they feel your personal security forces will create a problem, they won't hesitate to eliminate them immediately.

Each day of captivity is one of uncertainty. You will have no idea what will happen next. But history is on your side. Always keep in your mind that the likelihood of you coming through this ordeal without injury is high.

The actual move to seize you is the most dangerous of your ordeal. You must ask yourself whether or not you will resist the kidnap attempt. The attack will be over in a matter of seconds. Hans-Martin Schleyer's kidnapping took less than 90 seconds. Aldo Moro's kidnapping took approximately 45 seconds. The reason why they were over so quickly is because the terrorists were organized. More than likely the attack will catch you by surprise. Keep in mind that if you find yourself in the midst of a kidnapping it is because the terrorists have decided your defenses are weak, and that you are vulnerable. They have gone through their selection process and chose you as the victim. The decision to kidnap you was not made lightly; they have done their homework and feel that you can be taken easily. If you feel that you have a chance of escaping, take it, but let common sense be your guide. If the terrorist has a gun pointed at you, don't try anything that may provoke him into pulling the trigger. Even though kidnapping may be their original goal, don't provoke them into making it an assassination. No matter how reasonable they may appear, don't expect them to react rationally. The burning question in your mind while all this is going on is "Are they going to kill me?" The answer is probably no, at least not yet. If the kidnapping is for political reasons, you are a valuable propaganda tool. If the kidnapping is intended to gain concessions from your company, you are of no value to the terrorists dead.

What do you do when you are taken? At the moment of the kidnapping do you put up a fight or do you go quietly? If the kidnapping is a violent one, you will probably get hurt if you put up a fight. If you get through the actual kidnapping alive, you will probably go the rest of the way without serious injury.

The actual moment of kidnapping is a bad time for all involved. The kidnappers are nervous. They want to get it over with. They are very likely to overreact to any type of resistance. Your biggest concern at the moment of attack is whether it is a kidnapping or an assassination. If it's an assassination, you probably won't have to worry about it for long. Very seldom are executives ever assassinated outright; they are always more than likely kidnapped. At the precise moment of the attack, the situation will be very confusing. You won't be able to tell what's about to happen. The time frame is measured in seconds, and if the attackers have kidnapping in mind, they must take you alive. If you are considering escape, keep the following in mind:

- If the kidnapping takes place in a crowded area, the people around you will not come to your assistance. If the kidnappers have their weapons out, no one will be in a hurry to interfere and try to prevent the kidnapping.
- *Don't do anything foolish.* Your kidnappers will put you in a vehicle and drive you away. Leaping out of a moving car takes a lot of nerve, but it can also get you killed. Once you are in the vehicle and being driven away, you have lost. They have won. *Don't do anything foolish.*
- Your assailants will be armed. Past kidnappings have shown that during the actual kidnapping terrorists will not hesitate in using their weapons if things go wrong. Don't do anything that provokes them into escalating the violence.
- Remember that now is the time to think about what you're going to do and how you're going to do it.
- In an attack, you won't have time to think about what you're going to do. You will act through instinct. If you do some prior thinking about this unthinkable act, the chances increase that your instincts will be good ones.
- If you feel that you have the opportunity to escape, do it, but keep in mind the consequences of failure. If the odds are overwhelming, don't try to be a hero.

ESCAPE

Escape is something that you should think of, but not too much. There have not been many successful hostage escapes. If you decide to escape, timing is all-important. The best time to try to escape is while you are being abducted. Your next chance will come after you have been held in captivity for a period of time and a daily routine has been set by captors. During your abduction, you are closer to potential help than at any other time. You still know where you are and where you can go for help. Before you make your move, hold on. While this is the best time for escape, it is also the time for potential disaster. An escape attempt increases your chances for injury or worse. If you are going to escape, keep it simple.

If there are any actions you can take during the abduction phase that can safely prolong the process by two to three minutes, the kidnappers will more than likely give up. Their planning has put them on a precise time schedule. If the kidnapping starts to go beyond that time frame, they will abort the kidnapping. Remember that they don't want you, they want what you represent.

If you are thinking of escape, the emphasis is on the word *safety*. If you have the slightest doubt, do not attempt to flee. Instead, fully cooperate with your kidnappers.

Once the kidnappers have you they will take you to some safe place where they can hold you without being detected. They will have already done their research, and selected a location they are sure will be hard for police to find. The location they take you to is called a safe house. During this period of transport to the safe house, you can gain some valuable information that could aid you in the future. During the trip to the safe house try to stay alert and try very hard to be aware of your surroundings. The information you collect during this period of time could help find the kidnappers once you are released, or help if you try to escape. While you are in transit, try to:

- Identify the number, sex, and nationality of the kidnappers.
- Try to determine what sort of vehicle they put you in. Get the make and model if you can.
- Note the type of roads you are traveling on: Are they smooth, rough, do you feel you are on a highway or country road? Notice how fast the car is being driven. Are you stopping a lot? That would probably mean you're in a city. What can you hear? What can you see? It may not be much since kidnappers are fond of incarcerating their victims in windowless vans, and/or blindfolding their victims.
- Try to memorize every detail, no matter how minor. Try also to get an idea of the type of house that you're being taken to. Is it an apartment or farm house? Try to determine the direction you are traveling. During this stage don't do anything that aggravates the kidnappers. This is no time to be macho. Concentrate your entire effort on memorizing the details of the kidnapping and the kidnappers. This serves a two-fold purpose. First, you will be gathering potentially valuable information you can later use. Second, it will keep your mind busy and help control panic.

CAPTIVITY

Once in captivity, your greatest enemy will not be your kidnappers; it will be your attitude. Try to control your fear and despair. These two destructive emotions will quickly reduce your ability to resist and maintain your emotional stability.

Constantly remind yourself that your family, and company, are working for your release. The crisis management team will be working for your release within hours of your kidnapping. During confinement, your attitude towards many things will change. If you are not religious, you will become so. Remember the old war saying about how "there are no atheists in foxholes"? You're in the foxhole of your life right now. Religion and prayer will allow you not to be afraid of being afraid. Fear is natural, but don't let it paralyze you and turn you into an uncontrollable zombie.

Keep quiet. If you whimper and whine it may provoke more violence from your captors. Maintain your human dignity at all times, and have faith in the people trying to set you free.

The first few days will be the hardest. This is the time your captors will be feeling you out, trying to find your weaknesses. More than likely they will switch from being brutal to being kind. They may try to break you mentally. Keep in mind that this is the worst it will be and as time goes on the conditions of your captivity will probably get better. But be prepared to be humiliated. Humiliation is a powerful weapon when directed against people with great personal pride or those unaccustomed to public embarrassment. Stripping you of your clothing and forcing you to stand before an audience (a favorite terrorist tactic) can be devastating.

To help keep your morale and your sanity, attempt to maintain a measure of control of your environment. If you are held in isolation, develop and maintain a strict schedule. Sleep during a certain period, wake up at a certain time, do some isometric exercise on a regular basis. Give your days some structure. Do some mental work, think about the things that need to be done around the house and office. If there are problems at work that you had no time to solve, now you do.

U.S. Navy Commander Lloyd Bucher was the skipper of the Navy intelligence-gathering ship *Pueblo* when that ship was seized by North Korean air and naval forces on January 23, 1968. Bucher was kept in solitary confinement for nine of the twelve months he and the *Pueblo's* crew were held in North Korean jails. Psychological torture, brutal beatings, and conditions approaching sensory deprivation were among the barbaric punishments inflicted on Bucher. One of the mental techniques Bucher used to preserve his sanity was to attempt solving complex mathematical problems mentally. The attempt consumed large amounts of time, as did memory exercises he devised for himself, such as trying to recall every person Bucher had ever met in his life, starting with his earliest memories. According to Commander Bucher, both these techniques proven invaluable in saving his sanity during his captivity, the length of which he had no way of knowing.

A prisoner should eat everything he is given. The natural tendency to reject strange or foul-smelling food must be set aside. Self-imposed starvation does not improve a captive's ability to resist. Your "hosts" may go out of their way to make your food look bad. Make sure you eat, no matter how bad it looks or smells. Fill your day with any tasks available; go out of your way to think of things to do. Your goal is to make every activity last as long as possible and to have it monopolize your attention. At all costs, remain intellectually alert. Do mental exercises. Write a book about a subject you are familiar with. If you are a golfer, put together information on golf. Play imaginary rounds of golf. Go fly fishing in the Colorado Rockies. Do anything that keeps your mind working.

Time is something that all business executives are tied to. Suddenly, time, its restrictions and methods of measurement, have been taken away from you. You must develop some way of keeping track of time. It could be scratches on the wall, notches in a piece of wood, etc. Running your life according to time is a natural human habit and helps bring order to daily life. This will be taken away from you. It's vital to maintain some sort of schedule.

Be aware of psychological traps. There is a natural tendency to identify with the kidnapper's goals. You may begin to feel that payment of the ransom is all that stands between you and your freedom. Although this is probably true, do not give the kidnappers any help in deciding on a dollar value for the ransom. They could use this against your negotiators.

During your imprisonment you may have the opportunity to communicate with your family or the negotiators. This is where having a prearranged signal can be a benefit. Good preplanning will allow you to give your family signals that can help in finding your location. While in most occasions these signals are of marginal benefit, at the very minimum they can tell the negotiators of your general condition.

Once in captivity, be your own best friend. Your company, family, and friends will be making every attempt to set you free. Again, this is where having a crisis management team with a prearranged plan is helpful to you. You will know exactly what your team is planning and what they will be doing to secure your release. In captivity, your status has changed. You have gone from someone people rely on to a person relying on other people for your very existence. You must have faith in yourself. You are the same person you have always been, and you are probably much more intelligent than your captors. Use the intelligence, the street sense, the pride that have made you important enough to merit being a kidnap victim to stay alive and stay sane.

Your introduction to the kidnappers will be brutal, and you will naturally be concerned that they may continue with their brutality. You will be impressed by their apparent power. You will quickly come to realize your utter dependence on them. All this will come to affect you in ways you have never thought of.

It will be hard for you to control your own emotions. You will become despondent. But hang in there. Keep reminding yourself that there is a team of people working day and night to get you released.

If your captivity goes beyond a reasonable amount of time, adjust to your surroundings and try to make the best of a bad situation.

At all times retain your dignity; do not knuckle under to them under any conditions. More than likely they will be political extremists and will try to indoctrinate you into their ways. Listen intently and try not to aggravate them. Assert your dignity quietly and with firmness.

Learn as much about your captors as possible. Since they are all humans, there will be differences among them. You may be able to use all these differences to your advantage later. If you use your natural abilities, you will gradually be able to assert yourself in positive ways. Discovering the basic makeup of each individual, you will begin to establish a relationship that may keep your captors relaxed. Instead of being an object that they plan on making money with, you will become a person. As a person you may be able to take an active role in getting yourself out of this problem.

Always understand that there is someone trying desperately to help you while you are held in captivity. Within a day or so of your seizure the kidnappers will need to contact someone with their demands. You may be asked to cooperate. At this stage, the kidnappers' problem is *credibility*. They have to demonstrate that they have you in their power, alive and worth retrieving. They will take some article that belongs to you and send it to the negotiators, or they may have you talk into a tape recorder, photograph or videotape you. A common tactic to establish the date and time of the photo or video is to show you holding a dated copy of a newspaper. They have to show some proof that you still are in one piece. Don't fight them on this. In fact, be helpful. The more communication between your captors and the outside, the sooner you'll be released.

Once the demands have been received, the delicate process of negotiation begins. This is a point at which a personal crisis management program, one that you carry out yourself while being held, is absolutely needed. You can play a role as "advisor" to your kidnappers. If you had a crisis management program in place on the outside, you will know what demands will and will not be met by your people. Your captors may lie to you about the amount of money, and try to convince you that your company does not want to meet their demand. If you have a plan already in place on the outside, you already know the amount the company will pay.

There may be as many as three groups trying to gain your release: your family, professional negotiators hired by your company, and the authorities. There is potential for conflict among these groups. You are important to everybody, but whereas your family might give anything to get you back, the authorities have a wide responsibility and cannot give what rightfully belongs to others. If part of the demand is the release of prisoners held in their jails, that could mean long delays. Your organization, too, has its priorities and might find its position one of potential conflict. As part of their demands, terrorists may want ads denouncing the local government taken out in newspapers. This can create immense problems.

MEDIA: TOOL OR MENACE?

Let's talk about the media. Kidnapping is a noteworthy event. The media will not always be understanding, discriminating, or even aware of the harm it can do by letting out information. The media can be a help or a hindrance; it is rarely neutral. Representatives of the media can only help if their efforts are expertly guided to that end. To get a better understanding of the impact of the media on terrorism and kidnappings, look at the media's rules for dealing with the FBI in terrorist kidnap situations. The following guidelines on coverage of terrorists were issued on April 7, 1977, by the CBS news president, Richard Salant:

- News personnel should be mindful of the probable needs by the authorities who are dealing with the terrorists for communication by telephone and hence should endeavor to ascertain, wherever feasible, whether our own use of such lines would be likely to interfere with the authorities' communications.
- Responsible news representatives should endeavor to contact experts dealing with the hostage situation to determine whether they have any guidance on such questions as phraseology to be avoided, what kinds of questions or reports by established authorities on the scene should be carefully considered as guidance (but not as instruction) by news personnel.
- Local authorities should also be given the name or names of personnel whom they can contact should they need further guidance or wish to deal with such delicate questions as a newsman's call to the terrorists or other matters which might interfere with authorities dealing with the terrorists.
- Guidelines affecting our coverage of civil disturbances are also applicable here, especially those that relate to avoiding the use of inflammatory catchwords or phrases, the reporting of rumors, etc. As in the case of policy dealing with civil disturbances, in dealing with a hostage story, reporters should obey all police instructions but report immediately to their superiors any such instructions that seem to be intended to manage or suppress the news.
- Coverage of this kind of story should be in such overall balance as to length that it does not unduly crowd out other important news of the hour/day.
- Because the facts and circumstances of each case vary, there can be no specific self-executing rules for the handling of terrorist/hostage stories. News will continue to apply the normal tests of news judgment and if, as so often they are, these stories are newsworthy, we must continue to give them coverage despite the dangers of "contagion." The disadvantages of suppression are, among other things, (1) adversely affecting our credibility ("What else are the news people keeping from us?"); (2) giving free rein to sensationalized and erroneous word-of-mouth rumors; and (3) distorting our news judgments for some extra-

neous judgmental purpose. These disadvantages compel us to continue to provide coverage.

- Nevertheless, in providing for such coverage there must be thoughtful, conscientious care and restraint. Obviously, the story should not be sensationalized beyond the actual fact of its being sensational. We should exercise particular care in how we treat the terrorist/kidnapper. More specifically:
- An essential component of the story is the demands of the terrorist/kidnapper and we must report those demands. But we should avoid providing an excessive platform for the terrorist/kidnapper. Thus, unless such demands are succinctly stated and free of rhetoric and propaganda, it may be better to paraphrase the demands instead of presenting them directly through the voice or picture of the terrorist/kidnapper.

In his book *One American Must Die,* former TWA 847 hostage, Kurt Carlson comments on the excellent relations between his family and how CBS would not divulge that he was a U.S. Army Major. So it seems that CBS practices what they preach. The book is must reading for anyone who travels abroad.

UPI has established guidelines for the FBI on the reporting on acts of terrorism and kidnapping, saying that "There can be no clearly defined policy for terrorist and kidnapping stories since the circumstances vary in each story. However, we have established a set of guidelines that we think are workable in most circumstances."

The UPI guidelines:

- Each story will be judged on its own and if a story is newsworthy it will be reported despite the danger of contagion.
- Coverage will be thoughtful, conscientious, and show restraint.
- Stories will not be sensationalized beyond the fact of their being sensational.
- Demands of terrorists and kidnappers will be reported as an essential point of the story but not to provide an excessive platform for their demands.
- Nothing will be done to jeopardize lives.
- Reporters, photographers, and editors will not become a part of the story.
- If staffers do talk to a kidnapper or terrorist, they will not become a part of the negotiations.
- If there has been no mention of a deadline, kidnappers or terrorists will not be asked if there is one.

- In all cases, apply the rule of common sense.

TERRORIST OPTIONS

Once they have you in place, the terrorists have the following options:

1. Kill you because their demands are not being met;
2. Keep you alive hoping their demands will be met;
3. Negotiate with a willingness to settle for less than their full demands;
4. Free you and abandon the demands, perhaps surrendering in the process.

As time goes by, terrorists generally gravitate toward the latter options. If negotiations start, the process can be long and involved. You should realize that most first demands from terrorists are impossible to meet. The company will try to negotiate the sum downward. This is no reflection on your worth to the company. But it's somewhat like buying a car: The first price is not always the last price. When you're in captivity, it may seem like a bunch of strangers are playing with your life or it may seem foolish, but that's the way it will be. Be forewarned.

Taking an honest interest in members of the group holding you and showing gratitude for favors can help build a friendlier relationship. Arguments over political issues will only damage this bond and should be avoided.

Unexpected relationships can also develop during the holding phase of the kidnapping process. Living in a state of nearly total dependency, sharing intimate surroundings, and developing common concerns over potential assaults by third parties can result in the establishment of sympathetic ties between victims and captors. This phenomenon is commonly called the "Stockholm Syndrome" because a young female hostage in a Stockholm bank robbery actually fell in love with her captor. One of the victims interviewed explained the situation as being similar to a hospital patient's physical dependency on the hospital staff for all the necessities of life. This physical reliance sometimes evolves into a psychological dependency as well. An important part of the survival process is to recognize this phenomenon and to use it to your advantage.

Sooner or later every incident ends. Assuming a favorable outcome, the victim must be aware of some of the dangers and unique events which occur at the "point of release." Once again, the captors are highly vulnerable. They may be nervous, excited, and prone to overreact to any threatening behavior on the part of the victim or third parties.

If an assault or exchange of gunfire occurs, the victim must take advantage of any protection that is available. Keeping away from doors and windows and staying protected and under cover until ordered to do otherwise by law enforcement officials is essential.

Being rescued may not be the best thing that can happen to you. If the safe house where you are being held is located by the authorities, your situation can deteriorate dramatically. Your first concern in an assault is that it will be handled by the local police. They are more than likely not trained in hostage rescue. The first moments are the most dangerous, as the terrorists discover that they are now the ones surrounded and outnumbered. In a way, they are now the hostages. If they are not extremely rational people they may turn on you. If you have been able to develop a good relationship with them during your captivity, this will be less likely. Just before an assault, the police may tell them that they have no alternatives. They must either surrender or die. Once past this stage, if no firefight breaks out, negotiations will start. More than likely, one of the police officers is a trained negotiator. This can also cause you a great deal of anxiety, because the terrorists may make you their spokesman.

If all else fails, the police may attempt an assault on the safe house. There is nothing good to say about this tactic. Your only hope for survival is that the people doing the raiding know what they are doing. If they do, there are ingenious ways of effecting your release with little risk of harm coming to you. In countries where there is a great deal of terrorism, there are also police units that have made a science of hostage rescue. They can attack a safe house and have you out before you know what's happened. When living in a terrorist environment and well before your kidnapping, you should make discreet inquiries to ascertain if the police in that country have received hostage rescue training. If you are kidnapped it will be to your advantage (at least to the advantage of your mental health) to know if the people that may be rescuing you know what they are doing.

SKYJACKINGS

All the news that skyjackings receive tends to create some anxiety about flying. The likelihood of you being involved in a skyjacking is slim. Therefore you will probably never be held hostage aboard an airliner. But if you are, there's something you can do that will make the ordeal a bit easier and safer for you.

The most important thing not to do is argue with the terrorist. It's never a real smart idea to argue with someone holding a gun, especially if they're pointing it at you. There is no verbal argument that will effectively counter a pistol or a hand grenade.

If the hijackers start talking to you, answer them politely, in as calm a tone of voice as possible. When talking to hijackers—or if you are allowed to talk with other passengers—never whisper and never shout. Speak plainly, in a normal tone of voice.

Hijackers do not want people aboard trying to get a revolt going among the passengers. Whispering suggests conspiracy; shouting implies leadership as well as indignation. Don't do either.

The objectors, the plotters, and the wise guys will be among the first people sacrificed—not only to assure that they cause no more problems but also to serve as a warning that the hijackers mean business. The best move is to make no move at all. Do not become vocal . . . or a leader.

But it is wise not to attempt to get too friendly with hijackers of any type, either. In the process, the victim violates the first rule of survival by becoming conspicuous and notable. The Stockholm Syndrome—the sense of identification that grows between captor and captives—is well documented. The Stockholm Syndrome works to the terrorist's advantage—not to yours.

The psychotic hijacker generally cannot be reached safely by anyone but a trained hostage negotiator. Often this type of hijacker is paranoid or worse. Any approach to such a hijacker, no matter how friendly, may well be viewed as an attack.

If you are being held hostage by a criminal you will have nothing in common with them unless you want to talk about safecracking, how to outrun the cops, or the finer points of mugging. Otherwise they are not going to be responsive. One thing such a person does not respect or tolerate are signs of weakness. More than likely this person lives in a tough, antisocial world. They will not likely be converted or swayed by anyone on the plane, certainly not by a passenger who, as a hostage, lacks the stature of a proper authority figure.

As for political terrorists: Never try to con them. They read hijacking accounts and executive protection manuals like this one with more interest than 99 percent of their potential victims. The most knowledgeable terrorists not only have memorized the structure and layout of the plane, they know well the makeup of the human spirit and the workings of the human mind. Many are schooled beforehand on avoiding the weaknesses of the Stockholm Syndrome. They know what to look for in themselves and others, and how to avoid what's not needed for the success of the mission. Trying to play their heartstrings is not likely to get the passenger anything but a cursing and possible physical mistreatment, such as a painful beating.

Speak when spoken to by hijackers. Do as they say. Neither seek nor refuse favors. Face them when talking to them; do not look away. But do not try to face them down.

Be prepared for an unpleasant time—at best. And a long time at that, if the hijackers are terrorists. While hijackings by lone psychotics and criminals don't usually last as long, terrorist hijackings routinely last more than

24 hours . . . sometimes up to a week. During much of this time the plane will be on the ground, the air conditioning will be turned off and the doors will be closed. In particularly hot weather, the plane will become unbearable.

People will be hungry and thirsty. There will be nothing to give them. The toilets will overflow. Medical problems will abound—from nausea, headaches, and dysentery to coronaries. Some of the medical problems may well be caused by the hijackers. It is not unknown for hijackers to single out and physically mistreat a passenger, particularly if they feel thwarted or frustrated. It may be difficult to watch another passenger be beaten, pistol-whipped, or kicked. But trying to prevent it will not stop it. There will just be one more person injured . . . or killed.

Passengers and crew will have to ask permission to do the simplest, most basic things, like relieve themselves in front of others. Permission may be denied simply to demonstrate the total power hijackers have.

Besides the smell of unwashed, perspiring bodies and human waste, there will be a sickly odor of fear. There is no place to hide from the nightmare, no place for tranquility and privacy. The victim has to make his or her own tranquility.

One good way to do this is a form of meditation that, properly done, can be relaxing and even refreshing. The technique is simple: Breathe in. Count the breath and see the number. Exhale, then breathe in again. Count the breaths as you might sheep going over a fence. Concentrate on the number of the breaths. Visualize that number, not what is happening on the plane, not the fear, not anything but the number. Keep counting. When this is done properly, you will be aware of the surroundings but generally unaffected by them. Towards the end of the counting period, between ten and 30 minutes long, the breathing should automatically become slower. This method of mental escape and physical rest should be done two or three times a day.

Mental exercises such as counting to one million, writing letters or a story in your head, working on math problems, praying, or visualizing the route from home to office foot by foot helps pass the time. Isometrics and inobtrusive physical exercises, particularly those that flex the leg muscles to aid circulation, not only help relieve boredom, but keep you in good physical shape and ready to move quickly if you need to.

With luck, the hijacking will end calmly, with the passengers being released first and the gunman (or gunmen) surrendering. But sometimes, with some people, compromise is impossible. Then the decision may well be made to attack the hijackers before they can massacre the passengers. This type of decision will often be made when authorities detect signs that the decision has already been made to kill the hostages, e.g. when harsh, intransigent captors suddenly become ultra-pleasant to the captives while placing bombs or fire accelerants throughout the plane.

The technology and expertise is available to determine what is happening inside the aircraft and to pinpoint the location of the hijackers. Usually it is used if a counterterrorist attack is to be made. Then, at a time when those responsible for the situation deem best, the passengers and hijackers will go through hell.

If a decision is made to attack the plane and rescue the passengers, you will first be hit with a noise that stuns; light like an explosion in a blast will sweep the plane. These will probably be concussion grenades, intended to stun and otherwise incapacitate the terrorists. Smoke from small charges will drift through the plane; there will be hoarse yells and probably gunfire.

There is nothing to do but get down and stay down until the situation is sorted out. Preplanning the best way to dive out of the seat, or get the most protection from the seat, is a worthwhile mental exercise that can occupy some of the long hours before any hijacker/authority confrontation.

Do not attempt to help the authorities disarm or stop the hijackers. And do not stand up, even if the hijackers scream for you to do so. Anything or anyone that moves aggressively is likely to be hit by someone: Upward movement is aggressive; going for the floor is not.

When Israeli commandos pulled off their daring hostage rescue at Uganda's Entebbe Airport in 1976, the only passenger killed in the firing was one unfortunate young man who remained standing when the order came to get down. Get down and stay down.

Once freed, victims are often thrown back into a world they are not really ready for. They have gone through an experience that can have profound effects on their thinking. They see the world through eyes that have seen the effects of hijackings. All too often they are subjected to media people, long intelligence debriefings, and the welcome but emotionally charged task of reassuring relatives and friends by phone. And there is the rearranging of schedules and plans, and all the hassles of getting back underway in everyday life, for the world goes on just as if there had been no hijacking, no hostages.

What is important is to get cleaned up, get rest, and achieve a controlled release of your suppressed emotions, tensions, and fears.

14

Kidnap and Ransom Insurance

INTRODUCTION

If you or your company are doing business in a terrorist environment, you should carefully consider purchasing kidnap and ransom insurance (KRI). In today's terrorist-infested world, KRI is a necessity for anyone traveling in a foreign country. KRI has been the subject of some critical controversy in recent years. Some feel it guarantees a quicker payment of the ransom than would ordinarily take place, and therefore encourages kidnapping. This may or may not be true, but if you are the one being held captive there is no such thing as "quick payment of the ransom." The fact is that it can't be paid quickly enough.

KRI is an indemnification contract usually purchased by a corporation or wealthy individual seeking to recoup any losses resulting from ransom payments made to terrorists. The insurance company writing the KRI policy requires that the insured pay the ransom out of their own funds and then show proof of the payment to the insurance company for reimbursement. Details of KRI policies are closely guarded by the insurance companies, and rightfully so. Public disclosure of who has KRI and who doesn't could lead to a situation in which the terrorists pick only the insured as targets. The result would be a kidnapper's holiday, along with massive losses for insurance companies.

As a person who should consider a KRI policy, there are some basic facts you should know about this type of coverage.

THE HISTORY OF KRI

Lloyd's of London invented KRI following the abduction and murder of the Lindbergh baby in 1932. In the early years, Lloyd's dominated the industry in this type of coverage. Within the last decade, however, the insurance industry has changed dramatically, and other insurance underwriters offer this type of protection as well.

From KRI's humble beginnings it has grown to become a necessity in many parts of the world, and is something that any executive working overseas should have. Many multinational companies quietly supply their overseas staff with KRI. Of course, the actual number of companies is not known.

The cost of KRI is difficult to determine. In the early days of this type of coverage, a company with executives that made only infrequent trips abroad, and then only to relatively safe areas (Great Britain, Northern Europe), would have paid around $0.01-$0.03 per hundred dollars face value of insurance, or about $100-$300 for $1 million worth of maximum coverage for one executive per year. Increasing the number of people covered caused the premiums to rise, but at a somewhat less than proportionate rate. If the executives traveled to risky areas, the rate increased by three to ten times (to about $0.05-$0.30 per hundred). For an American executive who was a resident in a risky area, the rate rose 75-150 times the base rate, or to $1-$5 per hundred. Deductibles cut the costs considerably. One expert gave the following example: An Argentine subsidiary of an American multinational buying a policy with $5 million face value would pay about $2 per hundred for a single executive, or the same amount for four executives with $125,000 deductible.

The type of coverage changed in the 1980s. Insurance companies will now provide protection for all employees and their families. A U.S. corporation with ordinary foreign activities that has $100 million in sales could probably purchase a $5 million limit policy for $5,000 in annual premiums, and a $10 million limit for $7,500. These premiums are equivalent to $0.10 and $0.075 per hundred. Deductibles are used only in the most troubled areas and often there are no aggregate limits specified. For the dangerous areas of the world, however, the rates specified earlier for the narrow coverage policies are still applicable.*

*Brian Jenkins and Anthony Cooper, *Terrorism* (Stoneham, MA: Butterworth Publishers, 1978), pg. 195.

Premiums can and often do get expensive. Rates vary a great deal depending on who is covered and where they are located. Insuring everybody in the corporation, or everybody in a particular country, is very expensive compared to insuring the top officers. But insuring all of a company's employees working in an terrorist environment eliminates a serious problem that results in a company insures only a few of its top officers. How would you like to be the one who has to make the decisions as to who gets covered and how much the company is willing to pay to get them back? How would you like it if the fact that you're the one making these decisions became public knowledge within the company? These sorts of facts have a way of leaking their way into the rumor mill. How would you explain the selection process criteria to those who'd been left out? Wouldn't that be like saying that the company doesn't consider them worth insuring? That could leave you and the company open for lawsuits if that individual is kidnapped. To avoid this predicament, as well as concerns about uninsured employees' being abducted, strongly consider acquiring global coverage. What it costs in extra money, it pays for in peace of mind and good employee morale.

As with other types of insurance, KRI rates are based on the probability of claims being filed. If you are living in a country with a history of kidnappings, you will pay more for coverage than will those living in areas where kidnapping is less common. For example, if you were doing business in Bogotá, Colombia, or El Salvador, your company would pay much higher premiums than if you lived in Boston.

KRI premiums depend on a number of other factors too. Well-known, high-profile companies closely associated with the United States pay higher premiums. History shows that if your company is an overseas symbol of the U.S., it is a much more likely target than a company whose logo or name is not recognizable, or not closely associated with the U.S. If you work for an American oil or auto company or U.S. bank, your premiums will more than likely be higher because you are perceived by terrorists as very wealthy, and a principal promoter of multinational influence. The insurance company will take a close look at your lifestyle. If you insist on an ostentatious chauffeured vehicle (unless that chauffeured vehicle is an armored car, driven by a trained bodyguard) and live lavishly in a high-profile lifestyle, your rates will be higher than a similar executive who maintains a lower profile in the same country. KRI rates also depend on the nationality of the insured. Expatriates are considered higher risks than "native" executives, with Americans the highest of all: Americans are worth more in ransom; their presence in the insured group can drive the price way up.

Secrecy is a mandatory condition of kidnap/ransom insurance. If security is breached, the policy can be canceled and any ransom claims rejected. (No KRI policy, however, is ever known to have been canceled for that reason.)

A second policy condition requires that the insured corporation pay a ransom out of its own treasury, to be reimbursed as soon as the insuror is satisfied that the claim is legitimate.

KRI policies can be canceled if the insuror feels that a claim has been "mismanaged"—for example, if the insuror thinks the corporation is not being sufficiently hard-nosed in the ransom negotiations. When Argentine terrorists abducted one of a pair of brothers operating a large supermarket chain, the ransom was paid with such alarming speed and disregard for negotiation that, Lloyd's pointed out, the kidnappers would be encouraged to strike again. They did, carrying off the second brother. Again their demands were met immediately; but this time the insurance claim was refused.

KRI insurance is not viewed favorably by everybody. There are some countries that frown on KRI. Several European countries—most notably, Italy—prohibit or strongly discourage the writing of kidnap/ransom insurance. The argument against KRI is similar to that used against ransom itself: Some people and countries feel that it can actually encourage kidnappings, or that the existence of such a policy encourages the corporation to pay a greater ransom than it would otherwise be willing to part with. Since insurance companies continue to write KRI policies, it's evident that underwriters can still make money on such coverage, which is itself evidence that widespread kidnapping of KRI policy holders does not take place.

When an insurance company writes a KRI policy for a new client, it will usually offer to pay the cost of a security survey of the client's operations.

There are security survey companies recommended by insurance companies that will provide you not only with the surveys themselves, but will also negotiate your release if you are kidnapped. This works to your advantage. The insurance company will grant your company a discount on the insurance premium if the company does a survey. Security survey companies provide a service that will have one of their people on the scene of the kidnapping within 24 hours. This person will help with the negotiations, and in many cases, be a trained negotiator himself.

STANDARD POLICIES

A standard corporate kidnap and ransom/extortion insurance policy form covers:

- Kidnapping or alleged kidnapping
- Loss of ransom while the money is in transit
- Extortion threatening bodily injury
- Extortion threatening property damage
- Reward payments
- Miscellaneous expenses, such as judgments and salary continuation of the victim

- Loss of corporate and personal assets
- Product-recall expenses
- Government detention
- Expenses incurred due to business interruption
- Accidental death and dismemberment
- Consequential property loss due to extortion.

Policy coverage is as high as $50 million per occurrence and, depending on the policy, there may be no deductible. Policies can cover specific individuals or all corporation employees for a period of one or three years. Short-term policies also are available.

Rates are largely dependent on:

1. Degree of foreign exposure, including the number of employees and the countries in which they are located.
2. Size of the company, in terms of revenue and employees. The larger and the richer a company, the higher the visibility it is more likely to have.
3. Type of business. Businesses with political connotations have a higher exposure to terrorist activities.
4. Reputation and character of the insured.

COST/BENEFIT

A cost-benefit analysis of KRI is close to impossible to do. The underlying causes of loss are too volatile, and the terrorist climate can fluctuate greatly. An area of the world that may be safe one month can become a hotbed of terrorism six months later.

When you buy insurance you are trading a known cost for an unknown cost. You make the decision to buy the KRI based on whether your company feels that it can afford to lose say $10,000 in one year as a result of an insurance premium but decides it cannot afford to run the risk of a potential $10 million loss due to a ransom payment. Therefore, your company may decide to spend that $10,000 on an insurance policy for you that would cover a ransom demand of $10 million loss. It's a hypothetical situation, but it illustrates the type of thinking behind the decision to purchase this type of insurance coverage.

While there is no such thing as a typical ransom payment, demands of $10 million and more are not unusual. As indicated before, recent ransom demands have reached this level. A $10 million uninsured loss would be painful for any corporation, no matter what size.

No one likes to spend money on any form of protection in which the actual risk is not known; it's impossible to know if security measures have deterred the terrorist or not. If they don't strike, maybe your security did deter them, or maybe they never intended to strike. What real difference does it make? In either case, you're still here, safe and sound and not kidnapped. KRI is just another type of kidnapping defense, one that insures that your ransom will be paid promptly, your release secured, and your company's investment in you protected.

KRI COVERAGE

Premium costs and the scope of coverage afforded under each of the companies' policies is generally the basis for deciding with which company to place insurance.*

1. Business premises extension reimburses for any monies that must be brought in from outside for any kidnap situation if the money is lost while it is on the premises.
2. Transit extension reimburses for any monies that are stolen between leaving the premises and reaching the kidnappers.
3. Reward extension includes coverage for monies paid to informants whose information leads to the arrest and conviction of the individuals responsible for the kidnapping.
4. Personal assets extension reimburses the insured persons for their personal assets that are used as ransom payment if the demand is made on the insured person and not the corporation.
5. Negotiations, fees, and expenses reimburse for reasonable fees and expenses incurred to secure the release of a hostage, including interest on a bank loan to pay a ransom payment.
6. Property damage coverage extension provides coverage for threats to cause physical damage to property.
7. Defense costs, fees, and judgments cover costs resulting from any suit for damages brought by an insured person. (The importance of this extension is underlined by a case in which the kidnapped executive sued his employer for $185,000,000 in damages, claiming that the employer had not exerted sufficient efforts to free him and had not taken steps to protect him from such an occurrence after the company had been forewarned that the executive might be a target for possible terrorist abduction.)

There are some other general requirements, mostly concerning secrecy, which is a sticky point in KRI coverage. Again, the fact that you do have

*James F. Broder, *Risk Analysis and the Security Survey* (Stoneham, MA: Butterworth Publishers, 1984), pp. 81-82.

KRI protection should *never* be disclosed, that is, unless you happen to like being bound and gagged and thrown into the back of a windowless van and driven wildly through city streets with guns blazing . . . well, you get the idea.

Of course, your security department should be aware of who in the company has KRI coverage, but they're the only ones who should.

The specific details of each policy must be studied and adhered to in the event of a kidnap, otherwise the incident may be noninsurable. Some of these considerations are as follows:

- The ransom or extortion demand is specifically made against the named insured.
- The extortion demand is made during the time frame of coverage as set forth in the policy.
- The company is to take every reasonable precaution to insure that the knowledge of the existence of the coverage is not disclosed to anyone except senior officials of the corporation.
- If a kidnap occurs, every reasonable effort should be made to determine:
  1. That an insured person has been abducted (Note: Not all policies cover all employees).
  2. That the police and/or Federal Bureau of Investigation (FBI) have been notified prior to payment. Further, compliance with police and FBI instructions and recommendations, which are in the best interest of the victim, are accomplished to the extent possible.
  3. The insurance company is notified at the earliest practical time.
  4. The serial numbers of the ransom payment are recorded.

Many times the insuror will require the company to subscribe to a risk analysis survey. These risk analysis companies come in various varieties. Some are companies that provide a political risk analysis. One such company is Frost and Sullivan.

World Political Risk Forecasts provides assessments of political risk to international businesses in 70 countries. The forecasts evaluate risk for a variety of factors that can affect business operations, from expropriation and social turmoil to government policies on investments, trade, and currencies.

A network of 250 specialists provide the information for the political risk forecasts. Their reports, updated monthly, analyze the political environment, weighing the business consequences of changes in the domestic balance of power or external environment.

The political forecasts for each country are combined with economic data—some from public sources, other from country specialists—to produce country reports. These reports present an 18-month and five-year forecast of levels of risk for 13 factors affecting international business.

WHAT THE SERVICE PROVIDES

Subscribers to *World Political Risk Forecasts* receive all 70 country reports with their subscription organized by region in four large binders. These reports, averaging 50-60 pages, follow a standard format for easy reference and cross-country comparisons. Subjects covered are described below in "Format for Country Reports."

- Revised reports. Each country report is completely revised annually; five to seven of these revised reports are issued each month.
- Updates. Major events of the past month in countries covered are analyzed and any changes in the 18-month forecasts are indicated.
- F&S Political Risk letter. All updates are summarized and worldwide risk ratings for all 70 countries reviewed and revised.

Other companies produce data on terrorist events. There is a list of such companies that provide this service listed under Information Sources in this book. These companies also will do a risk analysis for your company. They will make recommendations that are usually prerequisites before a KRI policy will be issued to a company.

A

Residential Security Checklist

1. Has a security survey previously been accomplished on your residence? If so when and by whom? Were recommendations considered valid and were they adopted? If not, why not?
2. How long have you lived in your current residence?
3. Who lived in your residence prior to you and your family?
4. Was security a factor in the selection of your residence?
5. Were all locks changed or rekeyed after you moved in?
6. Have you established a strick control of all residential keys? List all persons having keys? Have you hidden keys outside your residence?
7. Do you receive mail at your residence?
8. Do you receive other deliveries at your residence, i.e., newspapers, groceries, milk, fuel oil, fresh water, etc? NOTE: Terrorists frequently use delivery or sales pitches to gain entrance to home for intelligence or attack purposes.
9. What procedures exist for accepting deliveries? Are suspicious or unexpected deliveries refused? Are identities of delivery persons checked and verified with appropriate dispatches prior to accepting delivery or allowing access?

10. What procedures have been established for checking out repairmen and service representatives? Are they asked for phone numbers of companies they represent and other identification that can be verified before they are allowed access? Are they closely supervised while in the residence? Do repairmen call by appointment only?
11. Have family members been cautioned not to unlock or open doors to persons they do not know and until identity is established?
12. Are doors and windows always kept locked?
13. Do you employ maids, yardmen or other servants? Did you advertise for them? Were they recommended by someone? If so, whom? Did they approach you and offer their services? Did you ask for references? Were references checked? Were police and security agencies checked for criminal records, terrorist or subversive connections?
14. Do servants live in? Do they have unrestricted access to the residence?
15. Have you recently fired any household servants, drivers or other employees? If so, whom and why? Have you had any quarrels or harsh words with servants?
16. Do you or family members provide transportation to and from work for servants? NOTE: This is a dangerous practice. Pay for taxi or bus transportation.
17. Do you know where your servants live? Has this been verified? How long have they lived there?
18. Do servants have access to your travel schedules? How far in advance do they know of vacations and changes in your schedule?
19. Do servants have access to keys? Are they allowed to remove keys from the premises?
20. Is your residence left unattended during the day? On a regular basis or occasionally?
21. Do you have residential guards? Are they American or foreign nationals?
22. Are they well trained in their duties? Are they qualified with weapons? Do they have power of arrest?
23. Have you ever experienced problems with your guards, i.e., sleeping on duty; not following established procedures; rudeness; gun play or carelessness? Was this properly noted to superiors? Was disciplinary action taken? Did it correct the problem?
24. Do you ever display the American flag at your residence? Have you removed your name from mailboxes, and other locations where passersby can observe it?
25. Do you have a local telephone? Is it reliable? Is your telephone number listed in a public directory? Base directory?

26. Have you received obscene, threatening, annoying phone calls or an unusual number of wrong numbers or silent callers? Have you maintained a log of such calls? Did you report calls to police or security? Were they investigated?

27. How do you and your family members answer the telephone? NOTE: Do not give out identity or other information until the identity of the caller is established.

28. Are you acquainted with your neighbors? Have you met to discuss mutual security matters?

29. Have neighbors reported any suspicious activity in the neighborhood or persons asking questions about you or your family? Have neighbors been asked to report such matters to you?

30. Are you and family members sufficiently familiar with neighborhood environment to notice strange persons or vehicles?

31. Are you and your family aware of locations of nearest police station, hospital, fire station and other sources of emergency help? Have you posted address and telephone numbers for emergency use?

32. Have you installed any alarm, locks or other security devices?

33. Is your home equipped with fire extinguishers? Are family members familiar with their location and use?

34. Have you obtained a first aid kit, flashlight, candles, battery powered radio and other emergency equipment? Are they regularly inspected?

35. Do you own a dog?

36. Do you keep weapons in your home? Are they secured? What kind do you have? Are family members familiar with their location and operations?

37. Do family members exercise care when discarding letters, personal papers and other materials to insure nothing is included in trash which could assist a terrorist in developing information about you? i.e., Envelopes containing name, rank, address; letters with personal information or information about travel plans; party maps showing residences of other Americans; phone books, orders, etc.

Office Checklist

1. A survey should be conducted to determine if the office is constructed in a manner to minimize unauthorized entry and to inhibit terrorist attack.
2. Coordination should be made with security or law enforcement personnel for recommendations on selection and use of protective security devices for offices. If possible, a high-perimeter fence or wall with lighting to inhibit entry should be added if it does not already exist.
3. If entry is normally restricted to organizational members, or restricted in periods of high stress, a peephole or small window aperture in doorways where visitors arrive should be available so that they can be observed for identification purposes; an intercom or interview grill will be helpful in this identification process.
4. Consider restricting entry to the interior of the office at all times; use a reception room for normal handling of visitors.
5. Consider, also, installing metal detection devices at controlled entrances; do not permit nonorganizational members to bring in boxes or parcels without inspection.
6. Make the offices as fireproof as possible. Ensure that fire extinguishers

are available and in working order; they should periodically be checked and recharged as required.

7. Emergency items such as a supply of fresh water, food, candles, lanterns, flashlights, extra batteries, blankets, portable radio, camping stove with spare fuel, ax, first-aid kit, and other appropriate items such as a bomb blanket should be maintained.
8. Select and prepare an interior "safe room" for use in case of attack.
9. Emergency exits, evacuation and escape routes should be selected, and the staff should be briefed on them.
10. Ensure that all members of the organization know the location of fire equipment, bomb blankets, fire escapes and other emergency exits, electrical service switches, weapons, and emergency radio.
11. Keep the grounds to the office well lit at night if appropriate. Exterior lighting is one of the least expensive and most effective deterrents to entry and cannot be overemphasized.
12. Be alert to persons disguised as public utility crews, road workers, vendors, etc., who station themselves near the office in a position to observe your activities and gather information for future attacks.
13. Determine if the local police patrol the area. If not, request that they do.
14. Consider using private guards.
15. Attempt to draw as little attention as possible to the location or purpose of your office. In high-risk areas, consider removing all identifying signs.
16. Ensure that each host country employee is given a security check; updating of earlier checks should be done periodically.
17. Ensure that all doors are locked at night, on weekends, and when the office is unattended; keep tight control of door and other keys.
18. Do not leave important or confidential papers unattended; ensure all are secured at all times.
19. Lock all cabinets and closets when not in actual use. Janitors should be under supervision when cleaning unless they have been cleared for security purposes—and even then they should be spot-checked.
20. Lock all office lavatories when not in actual use. They, too, should be cleaned only by, or under the observation of, a cleared employee.
21. Visitors should be escorted when visiting the office. Control of strangers seeking entrance should be maintained.
22. Do not discuss in the presence of or reveal to visitors the travel plans or timetables of staff or visits by members of other offices.
23. Notify the appropriate security personnel immediately of an unexplained package anywhere in or near the office; know what to do in bomb threat situations.

24. Do not reveal to telephone callers, unless sure who they are, the whereabouts of persons asked for.
25. There should be a standard operating procedure established for emergencies; and everyone must understand their role.
26. Lock the doors to service areas such as closets and cupboards.
27. Empty wastebaskets frequently to prevent an accumulation of trash which would hide a device.
28. Don't have parcels, boxes, cases, stacks or bundles of books or magazines in public areas.
29. Generally keep the premises clean and neat so that it will be easier to observe something out of place (both inside and around the exterior).
30. In rooms and offices, see if the construction of built-in furniture (or its placement) creates hiding places and alter, move, or change it to remove them.
31. If curtains, drapes, or cloth covers create hiding places, raise, tie back, or remove them.
32. Have any adjacent mailbox removed from proximity to the building.
33. Arrange for local officials to place a "No Parking" sign in front of the office entrance if possible.
34. Remove bicycle racks from near entrances or any wall of the office. Be alert for parked or abandoned bicycles or motorcycles near the entrance or walls; they may be bombs.
35. Fasten down or lock manhole covers near the building.
36. Lock desk drawers accessible to the public.

Personal Security Questionaire

1. What time do you leave for work?
2. Can you vary the hours you work?
3. Do you have regularly scheduled meetings each day, or week, or month? Can you change the time of the meetings?
4. Do you have lunch at the same place every day?
5. Do you have a general sports activity at a regularly scheduled time?
6. Are any of your leisure time activities conducted on a regular schedule?
7. How do you get back and forth to work? Private vehicle, chauffeur driven?
8. If you are in a chauffeur driven car has the driver been trained in evasive driving?
9. Do you change your routes to and from work?
10. Do you have a duress code for your driver to warn you when it is not safe to enter the vehicle?
11. Do you have a crisis management plan?
12. Do you use different entrances when entering and leaving work?
13. Do you know where to go to get help along your route to and from work?

14. Do you drive on well travelled roads?
15. Do you have a secured entrance to your office?
16. Is your parking space marked with your name and title?

D Responsibilities of the U.S. Principal Departments and Agencies

Numerous federal departments and agencies contribute to the national program to combat terrorism. The following provides a detailed listing of the various activities of those agencies with major responsibilities.

DEPARTMENT OF STATE

This department carries out programs for combatting terrorism in the following ways:

- Discharges its Lead Agency responsibilities for terrorism outside the United States.
- Maintains the security of U.S. overseas diplomatic and consular facilities.
- Cooperates with U.S. businesses as part of its effort to enhance the security of private U.S. citizens abroad.
- Conducts research and analysis on terrorism.
- Provides security for visiting foreign diplomats and dignitaries.
- Protects the Secretary of State.

- Provides training for personnel of U.S. overseas missions on security and crisis management.
- Provides antiterrorism training and assistance to civilian security forces of friendly governments.

The principal office involved in these functions are the Office of the Ambassador-at-Large for Counter-Terrorism; the newly created Bureau of Diplomatic Security; The Bureau of Intelligence and Research; the Office of Foreign Building Operations; the Foreign Service Institute; and the Office of Foreign Missions.

DEPARTMENT OF JUSTICE

The Department of Justice pursues the following counter-terrorism-related activities and programs through the FBI, the Justice Department's Criminal Division and the Immigration and Nutrualization Service:

- Carries out its Lead Agency function to prevent, respond to, and investigate violent criminal activities of international and domestic terrorist groups within U.S. jurisdiction.
- Investigates terrorist acts abroad under the new Hostage-Taking Statute that makes the hostage-taking of U.S. citizens overseas a federal crime.
- Collects and investigates intelligence on terrorists to predict potential movement or criminal activities.
- Investigates terrorist incidents and related criminal activities using investigative techniques to identify, arrest, prosecute, and incarcerate those responsible.
- Maintains operational liaison with local law enforcement agencies throughout the United States.
- Provides training in the field and at the FBI Academy in Quantico, Virginia.
- Participates with local and state authorities in joint terrorism task forces.
- Maintains contact with and conducts limited joint investigations with allied national police and security services on terrorism through 13 legal attache offices.
- Collects technical information regarding terrorist explosives and bombings within the United States and disseminates it to international bomb data centers.
- Heads the national Hostage Rescue Team, a special group of highly trained FBI agents who deal with critical terrorist situations.

- Provides legal direction and support during terrorism investigations.
- Supervises and coordinates subsequent prosecution of members of domestic and international terrorist groups whose acts violate federal criminal law.
- Inspects and determines eligibility for applicants to enter United States.
- Maintains national and local lookout systems containing data relating to excludable aliens, including suspect or known terrorists.

DEPARTMENT OF TRANSPORTATION

The Department of Transportation's Federal Aviation Administration, U.S. Coast Guard and the Office of the Secretary conduct antiterrorism programs by carrying out the following:

- Conducts Lead Agency responsibilities through the Federal Aviation Administration by promoting the security of civil aviation, including prevention of air piracy, sabotage, and criminal activities within the jurisdiction of the United States.
- Provides assistance to law enforcement agencies in interdicting movements into the United States of dangerous drugs and narcotics that may be connected with terrorist activities.
- Maintains operational, investigative, communications, and liaison arrangements, with many foreign governments and private organizations such as aircraft manufacturers' and airline pilots' associations.
- Devotes substantial resources to airport and aircraft security programs both inside the United States and abroad.
- Assures the safety and security of vessels, ports, and waterways, and their related shore facilities.
- Advises on transportation security matters; provides security programs to protect personnel, communications equipment, and facilities.

DEPARTMENT OF DEFENSE

Defense Department agencies involved in combatting terrorism include the National Security Agency, the Defense Intelligence Agency, and the Joint Chiefs of Staff. Individual armed services antiterrorist programs supplement the overall Defense Department effort. After the Iranian hostage rescue attempt, the Department of Defense established a counterterrorist organization with permanent staff and specialized forces. These forces, which report to the National Command Authorities through the Joint Chiefs of Staff, provide a range of response options designed to counter specific acts of terrorism.

Additionally, the Defense Department maintains worldwide technical collection systems for gathering round-the-clock information on terrorism, which it disseminates to other national counterterrorism efforts, maintains data on terrorist groups, and produces publications on incidents and advisory and warning messages.

CENTRAL INTELLIGENCE AGENCY

The Central Intelligence Agency and other elements of the intelligence community contribute vitally important intelligence to the NSC and the Lead Agencies before, during, and after terrorist incidents. The organization is particularly crucial in the flow of information between the United States and other countries.

Analytical units of the CIA prepare both current and long-term reports on terrorist organizations, individuals and trends, and disseminate these reports on a timely basis to all government agencies with counterterrorist responsibilities. Should the White House direct military action in a counterterrorist situation, the CIA is prepared to provide intelligence support to the Defense Department.

The Director of Central Intelligence has overall coordination responsibility within the intelligence community for counterterrorism. He has designated the National Intelligence Officer for Counterterrorism as the focal point to coordinate national counterterrorism intelligence activities and to ensure counterterrorism priorities are established for the intelligence community.

DEPARTMENT OF THE TREASURY

The Treasury's role in the fight against terrorism involves protection, investigation, intelligence, interdiction, training, and response. Its activities range from thwarting an assault on the President, to investigation of an arms export case, to imposing and enforcing economic sanctions on state sponsors of terrorism.

Principal Treasury agencies are the United States Secret Service, the Bureau of Alcohol, Tobacco, and Firearms, the United States Customs Service, the Federal Law Enforcement Training Center, and the Internal Revenue Service.

INTERAGENCY PROGRAMS

Numerous interagency bodies contribute significantly to the national program. The major coordination effort, however, is carried out by the Interdepartmental Group on Terrorism, which is chaired by the Department of State and comprised of representatives from over a dozen departments and agencies.

Specific working groups have been established under the auspices of the Interdepartmental Group on Terrorism. Some of the more noteworthy are:

- Technical Support Working Group—assures the development of appropriate counterterrorism technological efforts.
- Public Diplomacy Working Group—designed to generate greater global understanding of the threat of terrorism and efforts to resist it.
- Antiterrorist Assistance coordinating Committee—coordinates the antiterrorism training programs of State, Defense, and the CIA.
- Rewards Committee—develops procedures for the monetary rewards program for information on terrorists.
- Exercise Committee—coordinates antiterrorism exercise programs.
- Maritime Security Working Group—assesses port and shipping vulnerabilities to terrorism.
- Legislative Group—reviews legislative proposal and develops future antiterrorist initiatives.

Sources

These organizations supply security consulting services.

Inter-American Consultants
Steve Van Cleave
P.O. Box 30358
Atlanta, GA 72043

Ackerman & Palumbo
1666 Kennedy Caseway
Miami Beach, FL 33141

Control Risks Ltd.
4330 East West Hgwy.
Bethesda, MD 20814

M. Best Security Consultants Inc.
3881 Holl Way
Eagle, ID 83616

KIDNAP AND RANSOM INSURANCE

The companies below sell kidnap and ransom insurance.

Chubb Group of Companies
49 J.F. Kennedy Pkwy.
Short Hills, NJ 07078

Professional Idemnity Agency Inc. (PIA)
409 Manville Rd.
Pleasantville, NY 10576

TRAINING

Richard W. Kobetz
Arcadia Manor, Route 2, Box 100
Berryville, VA 22601
Conducts training on all
aspects of security

Police International Ltd.
Ron Newhouser
P.O. Box 220
Oakton, VA 22042
Courses in dignitary protection

Nueveidas International
Tony Cooper
P.O. Box 25571
Dallas, TX 75225
Courses in hostage survival
and corporate aircraft security

Lethal Force Institute
Massad Ayoob
P.O. Box 122
Concord, NH 03301
Shooting school

Scotti International
10 High Street
Medford, MA 02155
Courses in executive protection

BODYGUARD SERVICES

J. Mattman Security Inc.
645 South State College Blvd.
Fullerton, CA 92631

Vance International
10327 Democracy Lane
Fairfax, VA 22030

Information Sources

TERRORISM: AN ANNUAL SURVEY
J.L. Scherer 4900 18th Ave. S.
Minneapolis, MN 55417
Editor: J.L. Scherer
A chronological outline of terrorist incidents all over the world. Each incident has a brief description of the event. Also contains an incredible amount of statistical information on terrorism.

TVI JOURNAL
(Terrorism Violence Insurgency)
TVI Inc., PO Box 1055,
Beverly Hills, CA 90213
TVI is a scholarly journal on terrorism. Produced quarterly it is an excellent source of information on terrorism and managing the terrorist problem

PROTECTION OF ASSETS MANUAL
The Merritt Company
1661 Ninth St.,
Monica, CA 90404;
PO Box 955,
Santa Monica, CA 90406

A series of manuals on protection. A must for anyone who is responsible for protection.

SECURITY LETTER
Security Letter, Inc.
166 E 96th St.,
New York, NY 10128
Editor: Robert D. McCrie
An excellent newsletter on all aspects of security

AMERICAN SOCIETY FOR INDUSTRIAL SECURITY (ASIS)
1655 N. Ft. Myer Dr.,
Arlington, VA 22209
Officer: Ernest J. Criscuoli, Jr.
The biggest security association in the world. It contains seminars on security-related topics, and produces a monthly magazine on security.

ASSOCIATION OF POLITICAL RISK ANALYSIS
PO Box 1726, Grand Central Station
New York, NY 10163

NATIONAL SAFETY MANAGEMENT SOCIETY
6060 Duke St.,
Alexandria, VA, 22304
Officer: George E. Cranston, exec. dir.

RAND CORPORATION
1700 Main Street
Santa Monica, CA 90406-21

SUPERINTENDENT OF DOCUMENTS
U.S. GOVERNMENT PRINTING OFFICE
Washington, DC 20402

PUBLISHERS

Butterworth Publishers
80 Montvale Avenue
Stoneham, MA 02180

*They have a complete catalogue of books covering all aspects of security.

Paladin Press
P.O. Box 1307
Boulder, CO 80306

*A wide selection of hard to find books on personal security.

Rand Corporation
1700 Main Street
Santa Monica, CA 90406-2138

*Does research on all forms of terrorism.

MUST READING

Title: *Terrorism*
Author: Brian M. Jenkins
Publisher: Butterworth Publishers, Stoneham, MA 02180

*This is the personal security bible used by security professionals around the world. It may be to advanced for the average manager, but it is a good source of information.

Title: *In The Gravest Extremes*
Author: Massad Ayoob
Publisher: Massad F. and Dorothy A. Ayoob, Concord, NH 03301

*Anyone who owns a gun should read this book.

Publisher: *Living In Troubled Lands*
Author: Patrick Collins
Publisher: Paladin Press, Boulder, CO 80306

*If you plan on living in a foreign country, you need to read this book.

Title: *Risk Analysis And The Security Survey*
Author: James F. Broder
Publisher: Butterworth Publishers, Stoneham, MA 02180

*The title may mislead you. If you are responsible for your own security, it is a must to read.

Bibliography

Alexander, Yonah. *Terrorism.* Volume 7, Number 2. Crane, Russak & Company, Inc. 1984. 3 East 44th Street. New York, NY 10017.

Ayoob, Massad. *In The Gravest Extreme.* Massad F. and Dorothy A. Ayoob. 1980. 72 Broadway, PO Box 122. Concord, NH 03301.

Ayoob, Massad. *The Truth About Self Protection.* Bantam Books. 1983. 666 Fifth Avenue. New York, NY 10103.

Broder, James F. *Risk Analysis and the Security Survey.* Butterworth Publishers. 1984. 80 Montvale Avenue. Stoneham, MA 02180

Buckwalter, Art. *Surveillance & Undercover Investigation.* Butterworth Publishers. 1983. 80 Montvale Avenue. Stoneham, MA 02180

Clutterbuck, Richard. *Kidnap & Ransom.* Faber and Faber Limited. 1978. 3 Queen Square, London WC1N 3AU.

Collins, Patrick. *Living in Troubled Lands.* Paladin Press. 1981. PO Box 1307. Boulder, CO 80306.

Dobson, Christopher & Payne, Ronald. *The Terrorists.* Facts on File. 1979. 119 West 57th Street 1979. New York, NY 10019.

Friedberg, Ardy. *America Afraid.* Research & Forecasts, Inc. The New American Library Inc. 1983. New York, NY 10019.

Fuqua, Paul & Wilson, Jerry V. *Terrorism.* Gulf Publishing Company, Book Division. 1978. PO Box 2608. Houston, TX 77001.

JENKINS, BRIAN M. *Terrorism.* Butterworth Publishers. 1985. 80 Montvale Avenue. Stoneham, MA 02180.

JENKINS, BRIAN M. *TVI Journal.* Volume 6, Number 1. Editor and Chief. TVI Inc. 1985 Beverly Hill, CA 90123.

"Kidnapping: How To Avoid It, How To Survive It" Clandestine Tactics and Technology: Tactics And Countermeasures, (Gaithersburg, MD: The International Association of Chiefs of Police, Inc., 1979)

KOBETZ, RICHARD W. & COOPER, H.H.A. *Target Terrorism.* International Association of 1978. 11 Firstfield Road. Gaithersburg, MD 20760.

LINEBERRY, WILLIAM P. *The Struggle Against Terrorism.* Volume 49, Number 3. The H.W. Wilson Company. 1977.

SILJANDER, RAYMOND P. *Terrorist Attacks.* Publisher, Charles C. Thomas. Bannerstone House. 1980. 301-327 East Lawrence Avenue. Springfield, IL 62703.

SUPERINTENDENT OF DOCUMENTS. *Public Report Of The Vice President's Task Force On Combatting Terrorism.* 1986. U.S. Government Printing Office. Washington, D.C. 20402.

VARIOUS. *Contemporary Terrorism.* International Association of Chiefs of Police. 1978. 11 Firstfield Road. Gaithersburg, MD 20760.

Index

C

D

E

F

G

H

I

K

L

M

U

V

W

Executive Safety and International Terrorism

A Guide for Travellers